Impact of COVID-19 and Non-COVID-19 Vaccination in Special Populations

Impact of COVID-19 and Non-COVID-19 Vaccination in Special Populations

Guest Editor

Kay Choong See

MDPI

Basel • Beijing • Wuhan • Barcelona • Belgrade • Novi Sad • Cluj • Manchester

Guest Editor
Kay Choong See
Division of Respiratory and
Critical Care Medicine
National University Hospital
Singapore
Singapore

Editorial Office
MDPI AG
Grosspeteranlage 5
4052 Basel, Switzerland

This is a reprint of the Special Issue, published open access by the journal *Vaccines* (ISSN 2076-393X), freely accessible at: www.mdpi.com/journal/vaccines/special_issues/C22KZ5927S.

For citation purposes, cite each article independently as indicated on the article page online and using the guide below:

Lastname, A.A.; Lastname, B.B. Article Title. *Journal Name* **Year**, *Volume Number*, Page Range.

ISBN 978-3-7258-2646-9 (Hbk)
ISBN 978-3-7258-2645-2 (PDF)
https://doi.org/10.3390/books978-3-7258-2645-2

Contents

About the Editor

Kay Choong See

Adj. Assoc. Prof. Kay Choong See is currently a Senior Consultant Respiratory Physician and Intensivist at the National University Hospital, Singapore. He is also the Associate Designated Institutional Official for the National University Health System Residency Program. Adj. Assoc. Prof. See completed Advanced Specialty Training in both Respiratory Medicine and Intensive Care Medicine in Singapore, a Master's in Public Health at the Harvard School of Public Health, and a Master's in Health Professions Education at Maastricht University. He is competent in bronchoscopy, endobronchial ultrasound, ultrasound-guided pleural procedures, and thoracoscopy (pleuroscopy). His special interests include public health, medical education, and healthcare quality improvement.

Preface

The COVID-19 pandemic has highlighted the critical importance of vaccination and uncovered the complexities surrounding vaccination in various populations. This Special Issue, "Impact of COVID-19 and Non-COVID-19 Vaccination in Special Populations", aims to illuminate these nuances, to provide a comprehensive understanding of the barriers to vaccination and to advocate for evidence-based strategies that address these barriers. The scope of this collection encompasses a thorough examination of both COVID-19 and non-COVID-19 vaccinations, specifically targeting populations that exhibit varying degrees of vaccine hesitancy—namely children, adolescents, older adults, individuals with comorbidities, risk-averse individuals, and those with a history of COVID-19 infection. Through ten meticulously curated articles, the contributors address pivotal questions regarding vaccine efficacy, safety, and the reasons behind vaccine hesitancy, supported by robust clinical data and systematic reviews. We believe that through this endeavor, we can build trust and encourage higher vaccination rates among those who are most vulnerable to misinformation and worse health outcomes. This collection is intended for a diverse audience, including healthcare professionals, public health officials, researchers, and educators, all of whom play vital roles in expanding vaccination efforts. By disseminating these findings, we hope to promote a broader dialogue on the importance of vaccination and to foster collaboration among stakeholders aimed at achieving higher levels of immunization coverage. We extend our heartfelt gratitude to the authors who contributed their expertise and insights, making this Special Issue a valuable resource. Additionally, we acknowledge the support from our editorial team and peer reviewers, whose keen insights and rigorous examination of the submitted articles were instrumental in shaping this publication.

Kay Choong See
Guest Editor

vaccines

MDPI

Editorial

The Impact of COVID-19 and Non-COVID-19 Vaccinations in Special Populations

Kay Choong See

Division of Respiratory and Critical Care Medicine, Department of Medicine, National University Hospital, Singapore 119074, Singapore; kaychoongsee@nus.edu.sg

Vaccination to prevent human infection is a key driver for reducing morbidity and mortality. However, vaccine hesitancy, defined by the World Health Organization's Strategic Advisory Group of Experts on Immunization as "the delay in acceptance or refusal of vaccines despite availability of vaccination services" [1], can lead to under-vaccination. One of the reasons for vaccine hesitancy is the lack of confidence, or trust, in the efficacy and safety of vaccination.

To drive and sustain the vaccination effort and broaden the uptake by individuals, clinicians, and policy makers require evidence that demonstrates the efficacy and safety of vaccination not only to the general population, but also to special segments of the population that are prone to vaccine hesitancy. This Special Issue therefore aims to explore the positive and negative impact of COVID-19 and non-COVID-19 vaccination for these special populations. Ten papers have been published in this Special Issue that broadly address four population groups who are prone to vaccine hesitancy and contain information with important clinical and policy implications.

The first population group consists of individuals who are children [2], adolescents [3], and older adults [4]. Osman and colleagues conducted a test-negative matched case–control study among 14,161 children and adolescents aged 12–17 years in Qatar between 1 June and 30 November 2021 and demonstrated that a two-dose primary series of the Pfizer-BioNTech BNT162b2 mRNA COVID-19 vaccine provided a relatively high vaccine efficacy of 79%. A systematic review conducted by Tian and colleagues provides further data from 12 randomized controlled trials that support COVID-19 vaccination in children and adolescents under the age of 18 years by demonstrating high vaccine efficacy, high immunogenicity, and low rates of serious adverse events across various vaccine platforms. Ishak and colleagues focused on adults who are 75 years old and above and reviewed the vaccination recommendations in guidelines for the top 25 non-communicable diseases that are suffered by these older adults. The authors found that the current guidelines do not uniformly provide vaccination recommendations and generally omit information on the benefits and risks of vaccination, highlighting the need for guidelines that provide more comprehensive recommendations to promote vaccination uptake.

The second population group consists of patients with various comorbid conditions that could be perceived to blunt the efficacy of vaccination [5]. Widhani and colleagues performed a systematic review of COVID-19 vaccination in patients with autoimmune diseases. These patients are often immunocompromised from both their disease and from immunosuppressive medications. From the 76 studies included in their review, as expected, compared with healthy controls, patients with autoimmune diseases showed impaired immunogenicity to COVID-19 vaccines. The clinical impact of impaired immunogenicity differed between the vaccine platforms, with a 93% increased risk of breakthrough infections for inactivated vaccines and no increased risk for mRNA or adenovirus vector vaccines. Additionally, they found that a second dose of COVID-19 vaccination increased immunogenicity without elevating the risk of systemic adverse events. Ziemssen and colleagues arrived at similarly encouraging results for a very specific subgroup of

Citation: See, K.C. The Impact of COVID-19 and Non-COVID-19 Vaccinations in Special Populations. *Vaccines* **2024**, *12*, 79. https://doi.org/10.3390/vaccines12010079

Received: 31 December 2023
Accepted: 10 January 2024
Published: 12 January 2024

23 patients who received ofatumumab—a human anti-CD20 monoclonal antibody—for relapsing multiple sclerosis and reported good seroconversion rates for the COVID-19 mRNA booster vaccination.

The third population group consists of risk-averse individuals who perceive vaccination-related adverse events to be common or serious [6]. These individuals include those with prior severe drug-, food-, or insect sting-related allergic reactions like anaphylaxis, who were found by Asperti and colleagues to be more anxious when receiving COVID-19 vaccination compared with those with a mild allergy. Such anxiety was lowered by having the vaccination administered in dedicated facilities while supervised by an allergist. Other less acute but serious adverse events might worry some individuals. One of these adverse events is sensorineural hearing loss, which Liew and colleagues studied in their systematic review. The incidence of post-vaccination sensorineural hearing loss was fortunately very low at 0.6–60.77 cases per 100,000 person years for both COVID-19 and non-COVID-19 vaccines, which was comparable to the incidence of all-cause hearing loss, suggesting no excess risk from vaccination. Given the uncertainty about the increased risk of side effects with repeated vaccination, Soegiarto and colleagues studied 75 healthcare workers in Indonesia who received a third dose of heterologous COVID-19 mRNA booster vaccine after a two-dose series of inactivated vaccines. They found that the mRNA vaccination elicited a more robust antibody response compared with a third dose of inactivated vaccine, with minimal systemic side effects. Further evidence of the safety of three vaccine doses comes from an online survey in Saudi Arabia, conducted by Aldali and colleagues. Among 413 participants in the general population, individuals mostly reported mild to moderate side effects lasting less than four days after a three-dose series of various COVID-19 vaccines.

The fourth and final population group consists of individuals who have been previously infected [7]. Qin and colleagues found that nearly 60% of people who have recovered from COVID-19 infection experienced pandemic fatigue, defined by the World Health Organization as the "natural and expected reaction to sustained and unresolved adversity in people's lives" [8]. As pandemic fatigue has been linked to vaccine hesitancy, this study highlights the need to especially educate and encourage this population segment to receive further vaccination.

In conclusion, the papers in this Special Issue provide good support for vaccination to prevent disease and preserve health. Given the timing of this Special Issue in 2023, which coincided with the COVID-19 pandemic, most papers unsurprisingly involved COVID-19 vaccination. Nonetheless, the contributing authors have provided information that can be generalized to non-COVID-19 vaccination. Addressing the concerns of special populations at both ends of the age spectrum, patients with immunocompromising comorbid conditions, risk-averse individuals, and individuals experiencing pandemic fatigue can then help realize the full value of vaccination to maintain good health, safeguard economic activity, and avoid large-scale societal disruptions like pandemic lockdowns and border closures.

Conflicts of Interest: The author declares honoraria for talks sponsored by AstraZeneca plc, GSK Inc. Moderna Inc., and Pfizer Inc.

References

1. World Health Organization. Report of the SAGE Working Group on Vaccine Hesitancy. Available online: https://cdn.who.int/media/docs/default-source/immunization/sage/2014/october/sage-working-group-revised-report-vaccine-hesitancy.pdf?sfvrsn=240a7c1c_4 (accessed on 21 October 2023).
2. Cataldi, J.R.; O'Leary, S.T. Parental Vaccine Hesitancy: Scope, Causes, and Potential Responses. *Curr. Opin. Infect. Dis.* **2021**, *34*, 519–526. [CrossRef]
3. English, A.; Middleman, A.B. Adolescents, Young Adults, and Vaccine Hesitancy: Who and What Drives the Decision to Vaccinate? *Pediatr. Clin. N. Am.* **2023**, *70*, 283–295. [CrossRef] [PubMed]
4. Nicholls, L.A.B.; Gallant, A.J.; Cogan, N.; Rasmussen, S.; Young, D.; Williams, L. Older Adults' Vaccine Hesitancy: Psychosocial Factors Associated with Influenza, Pneumococcal, and Shingles Vaccine Uptake. *Vaccine* **2021**, *39*, 3520–3527. [CrossRef] [PubMed]
5. Tsai, R.; Hervey, J.; Hoffman, K.; Wood, J.; Johnson, J.; Deighton, D.; Clermont, D.; Loew, B.; Goldberg, S.L. COVID-19 Vaccine Hesitancy and Acceptance among Individuals with Cancer, Autoimmune Diseases, or Other Serious Comorbid Conditions: Cross-sectional, Internet-Based Survey. *JMIR Public Health Surveill.* **2022**, *8*, e29872. [CrossRef] [PubMed]

6. Lepinteur, A.; Borga, L.G.; Clark, A.E.; Vogele, C.; D'Ambrosio, C. Risk Aversion and COVID-19 Vaccine Hesitancy. *Health Econ.* **2023**, *32*, 1659–1669. [CrossRef] [PubMed]
7. Do, D.P.; Frank, R. Prior COVID-19 Infection: An Underappreciated Factor in Vaccine Hesitancy in the USA. *J. Public Health* **2022**, *44*, 471–474. [CrossRef] [PubMed]
8. World Health Organization. Available online: https://www.who.int/europe/news/item/07-10-2020-how-to-counter-pandemic-fatigue-and-refresh-public-commitment-to-covid-19-prevention-measures (accessed on 31 December 2023).

vaccines

Article

Pfizer-BioNTech mRNA Vaccine Protection among Children and Adolescents Aged 12–17 Years against COVID-19 Infection in Qatar

Khadieja Osman [1], Jesha Mundodan [1,2,*], Juel Chowdhury [1,2], Rejoice Ravi [1,2], Rekayahouda Baaboura [1,2], Jeevan Albuquerque [1,2], Bilal Riaz [1,2], Reem Yusuf Emran [1,2], Khatija Batoul [1], Abdul Mahmood Esameldin [1], Zinah Al Tabatabaee [1], Hayat Khogali [1,2] and Soha Albayat [1,2]

[1] National COVID-19 Track & Trace Team, Ministry of Public Health, Doha P.O. Box 42, Qatar; dr.khadijahassan@gmail.com (K.O.); juel.chowdhry@moph.gov.qa (J.C.); rravi@moph.gov.qa (R.R.); rekayahouda.baaboura@moph.gov.qa (R.B.); jalbuquerque@moph.gov.qa (J.A.); briaz@moph.gov.qa (B.R.); remran@moph.gov.qa (R.Y.E.); hkhogali@moph.gov.qa (H.K.); salbayat@moph.gov.qa (S.A.)
[2] Vaccination Section, Ministry of Public Health, Doha P.O. Box 42, Qatar
* Correspondence: jmundodan@moph.gov.qa

Abstract: Qatar was also hit hard by the global pandemic of SARS-CoV-2, with the original virus, Alpha variant, Beta variant, Omicron BA.1 and BA.2 variants, Omicron BA.4 and BA.5 variants, and Delta variant, sequentially. The two-dose primary series of BNT162b2 (Pfizer-BioNTech) COVID-19 vaccine against SARS-CoV-2 infection has been approved for use in 30 μg formulations among children and adolescents aged 12–17 years as of 16 May 2021. This study aimed at estimating the effectiveness of the 30 μg BNT162b2 Pfizer-BioNTech mRNA COVID-19 vaccine against the pre-Omicron variants of SARS-CoV-2 infection in children and adolescents aged 12–17 years residing in Qatar. A test-negative matched case-control study was conducted. The subjects included any child or adolescent aged 12–17 years who had been tested for SARS-CoV-2 using RT-PCR tests performed on nasopharyngeal or oropharyngeal swabs, as part of contact tracing, between June and November 2021, and was eligible to receive the BNT162b2 vaccine as per the national guidelines. Data regarding 14,161 children/adolescents meeting inclusion–exclusion criteria were retrieved from the national Surveillance and Vaccine Electronic System (SAVES). Of the total, 3.1% (444) were positive for SARS-CoV-2. More than half (55.96%) were vaccinated with two doses of Pfizer-BioNTech-mRNA COVID-19 vaccine. Amongst those immunized with two doses, 1.2% tested positive for SARS-CoV-2, while 5.6% amongst the unvaccinated tested positive. The vaccine effectiveness was calculated to be 79%. Pfizer-BioNTech mRNA COVID-19 vaccine provides protection from COVID-19 infection for children/adolescents; hence, it is crucial to ensure they receive the recommended vaccines.

Keywords: adolescent; children; 12–17 years COVID-19 vaccine; fully vaccinated; partially vaccinated; vaccination status; RT-PCR-positive; Pfizer; BNT162b2; vaccine effectiveness

Citation: Osman, K.; Mundodan, J.; Chowdhury, J.; Ravi, R.; Baaboura, R.; Albuquerque, J.; Riaz, B.; Emran, R.Y.; Batoul, K.; Esameldin, A.M.; et al. Pfizer-BioNTech mRNA Vaccine Protection among Children and Adolescents Aged 12–17 Years against COVID-19 Infection in Qatar. *Vaccines* **2023**, *11*, 1522. https://doi.org/10.3390/vaccines11101522

Academic Editor: Kay Choong See

Received: 12 July 2023
Revised: 3 September 2023
Accepted: 18 September 2023
Published: 25 September 2023

1. Introduction

The COVID-19 pandemic was an unprecedented health emergency and could be controlled to an extent by a stable public health system through early containment by early detection of infected persons, the isolation of infected cases, as well as contact tracing, testing, and quarantine of these contacts [1]. In addition, nonpharmaceutical preventive health measures, such as hand washing, using face masks, physical distancing, stay-at-home orders, school and venue closures, workplace restrictions, and environmental cleaning, were adopted internationally. These responses were modified, changed, or intensified with the emergence of new epidemiologic data, experience-sharing from other countries, and emerging newer variants.

Qatar experienced five waves of SARS-CoV-2 infection, by the index virus [2], the B.1.1.7 (Alpha) variant [3], the B.1.351 (Beta) variant [4], the B.1.1.529 (Omicron) subvariants

BA.1 and BA.2 [5], the Omicron subvariants BA.4 and BA.5 [6], and a prolonged phase of the B.1.617.2 (Delta) variant [7], sequentially. Community transmission of the SARS-CoV-2 Delta variant (B.1.617.2) was first identified in Qatar by the end of March 2021 [2,8,9]. Although Delta variant incidence increased and reached about 200 cases per day in the summer of 2021, it remained low compared to earlier variants incidences. Between 23 March 2021 and 7 September 2021, 43.0% of diagnosed infections were Delta variant infections [2,4]. The first Omicron variant infection in Qatar was identified on 24 November 2021. Within four weeks, it became the predominant strain [10].

Free SARS-CoV-2 testing was widely available in Qatar and was required for those in close contact with an infected person, with symptoms such as fever or acute respiratory illness, as well as people returning from travel abroad. All specimens collected via nasopharyngeal/oropharyngeal swab, irrespective of where they were collected, be it private or public health facilities, were tested at the National Virology laboratory, using real-time PCR tests, following the national testing standards.

However, vaccination remains the most efficient and effective control strategy against COVID-19. Global vaccine development efforts have been accelerated in response to the devastating COVID-19 pandemic, like accelerated evaluation procedures and authorization for emergency use. Several pharmaceutical companies were trying to develop an effective vaccine against severe acute respiratory syndrome coronavirus 2 (SARS-CoV-2) infection. Phase III trials reported high vaccine effectiveness (VE) against SARS-CoV-2 infection, with 70.4% for Oxford-AstraZeneca vaccine (ChAdOx1 nCoV-19 vaccine) [4], 95.0% effectiveness with Pfizer BioNTech vaccine (BNT162b2 mRNA vaccine) [5], and 94.1% with Moderna vaccine (mRNA-1273 vaccine) [6].

Qatar was among the first Gulf Cooperation Council (GCC) countries to procure COVID-19 vaccines and start the COVID-19 vaccination campaign nationwide for the citizens and residents. The COVID-19 vaccination campaign plan was developed by the National Strategic Committee and implemented by the Health Protection and Communicable Diseases Department (HP-CDC) of the Ministry of Public Health (MOPH) along with Hamad Medical Corporation (HMC) and Primary Health Care Corporation (PHCC). Four COVID-19 vaccines, namely, Pfizer, Moderna, AstraZeneca, and Jansen & Jansen, were approved in Qatar. A few others, like Sinopharm, Sputnik, and Sinovac, were conditionally approved. Guidelines and recommendation for COVID vaccines were prepared and distributed and are also regularly updated with the development of new scientific evidence globally.

In Qatar, vaccination commenced on 23 December 2020, primarily with the Pfizer BNT162b2 mRNA vaccine. The Moderna mRNA-1273 vaccine was introduced later. Vaccines were provided free of cost to all nationals and residents of Qatar through the public health care system and mass vaccination centers such as Qatar National Convention Centre (QNCC), VCIA (Vaccination Center Industrial Area), and Qatar Vaccination Centre (QVC). The primary focus or target groups were the high-risk groups, namely, frontline health care workers, those with chronic illnesses, and the elderly population. Once almost half the primary targets were covered, other categories like teachers and the workers living in close proximities and dormitories were focused through VCIA and QNCC. The mass vaccination campaigns for the general public helped to effectively increase the vaccine coverage for COVID-19 in Qatar, which helped reduce the Delta variant transmission in the community.

The highlight was that Qatar provided free vaccination for all. All entities providing vaccine were mandated to enter the vaccination details in the national vaccine registry (SAVES). A well-established Adverse Event Following Immunization (AEFI) reporting platform existed for the clinicians to report any suspected AEFI. In addition to this reporting platform for physicians, a link was provided on the ministry's website so that any individual experiencing any adverse event following COVID vaccination could register, and these data was analyzed monthly. Hence, it was easy to track adverse events following COVID vaccine administration. Only eight AEFIs were reported among those 12–17 years old post-vaccination with Pfizer BioNTech vaccine, of which five were related to the first dose. Most of the reported adverse events were mild reactions. Two of them were severe reactions—one was a case of myocarditis and the other anaphylactic reaction—both following the second dose, and these cases were hospitalized. However, there no deaths reported following vaccination. Hence, the Pfizer BioNTech vaccine in this age group was deemed safe, which, in turn, increased the uptake.

More than 80% of the resident population in Qatar had completed the primary series with either BNT162b2 (Pfizer-BioNTech) vaccine or the mRNA-1273 (Moderna) vaccine by September 2021 [5–7]. Pfizer-BioNTech (BNT162b2) and Moderna (mRNA-1273) mRNA-based vaccines are given as two doses scheduled three to four weeks apart, keeping a minimum of 15 days between the two doses.

As the vaccination was scaled up, the country faced two back-to-back waves of SARS-CoV-2 from January 2021 to June 2021, which predominantly were B.1.1.7 (Alpha) and B.1.351 (Beta) variants [6,9,11]. Community transmission of the B.1.617.2 (Delta) variant was first detected towards the end of March 2021, and it had become the dominant strain circulating by the summer [12–14].

While children tend to experience less symptomatic SARS-CoV-2 infection than adults, it is important to note that schools, youth sports, and other community events can still contribute to outbreaks and transmission. These settings can pose a significant risk even with high adult immunization rates [15]. The absence of in-person learning during the pandemic has had a detrimental impact on children. Given the vaccine's favorable safety and side-effect profile, high efficacy, and acceptable risk-to-benefit ratio in adolescents, evaluating its effectiveness in younger age groups is justified. Vaccinating adolescents can enable them to safely return to in-person learning and reintegrate into society, addressing the debilitating mental health consequences of the COVID-19 pandemic [16,17]. Reducing COVID-19-related morbidity and mortality in adolescents can be achieved by administering a safe and effective vaccine. Additionally, the availability of effective vaccines for adolescents is crucial in decreasing the reservoir of SARS-CoV-2. In line with this, the BNT162b2 vaccine has received emergency use authorization for adolescents aged 12 to 15 [18].

The World Health Organization's (WHO) Strategic Advisory Group of Experts (SAGE) on immunization updated the roadmap for prioritizing COVID-19 vaccines on 21 January 2021. Children and adolescents with comorbidities were identified as medium-priority population groups for administering the primary series and booster doses. In contrast, healthy children and adolescents were identified as the low-priority use group because of their low risk of severe disease, hospitalization, and fatality. European Union countries recommend primary vaccination against COVID-19 for 12–17-year-olds [7].

Ministry of Public Health (MOPH) Qatar approved Pfizer-BioNTech (BNT162b2) COVID-19 vaccine administration to children and adolescents, based on regional and global studies showing its safety and efficacy in this age group [8,9]. The two-dose primary series of BNT162b2 (Pfizer-BioNTech) COVID-19 vaccine against SARS-CoV-2 infection has been approved for use in 30 μg formulations among children and adolescents aged 12–17 years as of 16 May 2021 and 10 μg formulations among children aged 5–11 years as of 30 January 2022.

The objective of this study was to assess the effectiveness of the 30 μg dose of BNT162b2 COVID-19 vaccine (Pfizer) against severe acute respiratory syndrome coronavirus 2 (SARS-CoV-2) infection among children and adolescents aged 12–17 years, in Qatar, before the emergence of the Omicron variant.

2. Materials and Methods

Study design: a test-negative matched case-control study design [11,19,20].

Study population: children and adolescents aged 12–17 years, residing in Qatar, who had undergone COVID-19 tests using reverse transcription polymerase chain reaction (RT-PCR) for SARS-CoV-2 on nasopharyngeal swabs (NPS) or oropharyngeal swabs (OPS) as part of contact tracing, between 1 June and 30 November 2021. This ensured that the first batch of children/adolescents would have received both doses of the primary series and excluded Omicron variant-positive cases from the analysis. RT-PCR testing detects SARS-CoV-2 RNA at low levels, with analytic sensitivity of (98%) and specificity of (97%) [21].

Cases: children/adolescents aged 12–17 years with positive test results on RT-PCR for SARS-CoV-2.

Controls: children/adolescents aged 12–17 years who had negative test results on RT-PCR for SARS-CoV-2, matched by calendar week for the RT-PCR tests.

2.1. Inclusion Criteria

- Any children and adolescents aged 12–17 years residing in Qatar tested for SARS-CoV-2 using RT-PCR between 1 June 2021 and 30 November 2021 were eligible irrespective of nationality, gender, and vaccination status.
- Children and adolescents eligible to receive the Pfizer-BioNTech mRNA COVID-19 as per Ministry of Public Health (MOPH) guidelines.

2.2. Exclusion Criteria

- Children and adolescents tested for COVID-19 using a method other than RT-PCR.
- Uncertainty about the COVID-19 test results, which includes "Inconclusive" results or if results were unavailable for any reason.

Sampling: A total of 14,298 children and adolescents aged 12–17 years who were tested for COVID-19 in the date range of the study, vaccinated or unvaccinated, were extracted from the national database. Of this total, 137 had inconclusive test results and, hence, were removed, leaving us with 14,161 children and adolescents between 12 to 17 years, irrespective of vaccination status, who were included in the study. The study population selection process is illustrated in Figure 1.

The effect modification of the vaccine effectiveness by differences in the variants exposed to, changes in testing frequency over time, and differences in infection-derived immunity among the unvaccinated were adjusted by matching the cases and controls by calendar week for the RT-PCR test [22–25]. It was possible to find PCR-negative matches for most age groups due to the higher number of PCR-negative test results than PCR-positive results. Nonpharmaceutical interventions (NPIs), including face masks, social distancing, hand washing, and hand hygiene, were mandated in Qatar during the study period, with varying levels of restrictions for the public as per MOPH guidelines; hence excluding the confounding effect expected due to change in behavior after vaccination. Cases and controls were matched in a 1:5 ratio to maximize statistical power.

Data collection: COVID-19 case investigation teams receive a list of laboratory-confirmed SARS-CoV-2 cases from various government and private sectors nationwide. Team members contact the index case by phone to obtain the necessary details and record the number and details of those the patient had close contact with in the last 48 h. A close contact refers to anyone who has met someone infected with the COVID-19 virus, starting from 2 days before the onset of the infected person's illness up to 14 days after. Based on this information, a swabbing dispatch list was prepared daily, which is then forwarded to the field team supervisor for action. The swab team successfully visited the homes and workplaces of these confirmed cases and collected the necessary swabs from the close contacts enlisted. It was recommended to collect nasopharyngeal and oropharyngeal swabs in a single vial containing transport medium and only oropharyngeal swabs for children under 14 years of age.

Nasopharyngeal and oropharyngeal swabs collected across the country (irrespective of public/private), including those by the field teams, were tested using the RT-PCR tests for SARS-CoV-2 at the National Virology Laboratory under Hamad Medical Cooperation (HMC). The MOPH database, Surveillance and Vaccine Electronic System (SAVES), receives all real-time RT-PCR test results from the laboratory [10,26]. The data regarding COVID-19 laboratory testing, vaccination (which includes the types of vaccine and dates of the first and second doses of vaccine administration, place of immunization, expiry date of the vaccine, and the lot number), and associated demographic information were retrieved from the national integrated digital health information platform, Surveillance and Vaccine Electronic System (SAVES), owned by the Ministry of Public Health (MOPH), Qatar. The vaccination details of the citizens, residents, and visitors who had been vaccinated abroad were incorporated into the National Vaccine registry upon arrival in Qatar [27].

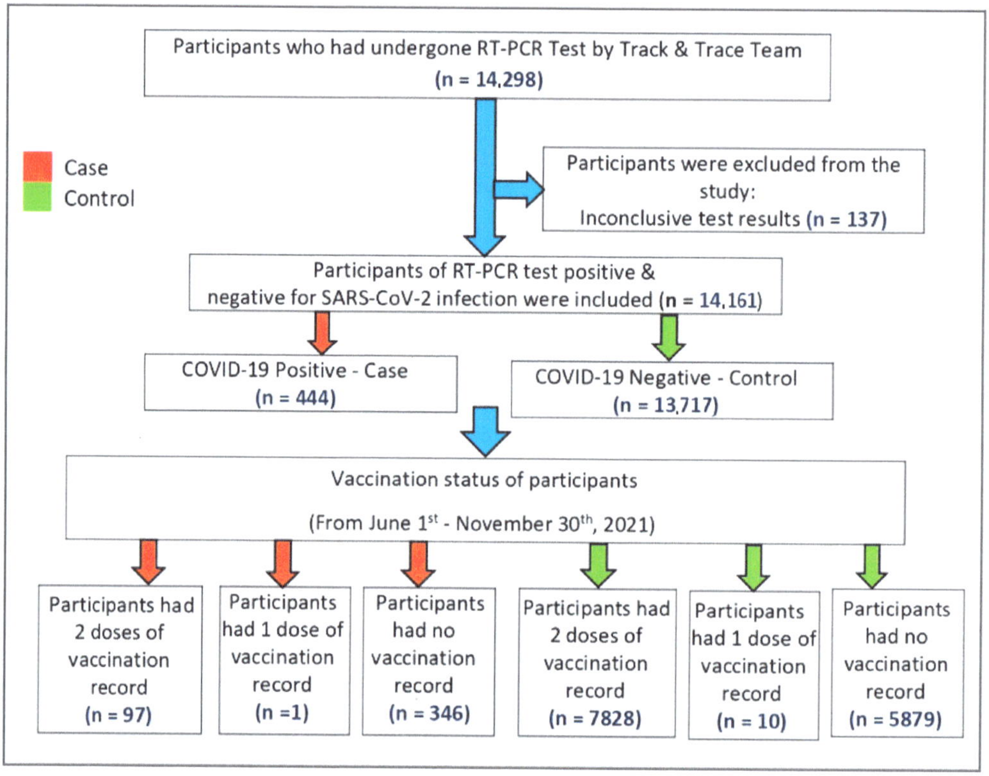

Figure 1. Flowchart illustrating the selection process of participants for investigating Pfizer-BioNTech mRNA vaccine effectiveness against SARS-CoV-2 infection.

The study participants were divided into 3 categories based on vaccination status: fully vaccinated and immune (those who had completed 14 days after receiving the second dose of vaccine), partially vaccinated or partially immune (participants who had received only 1 dose of vaccine or those who had not completed 14 days after receiving the second dose of vaccine) and unvaccinated (participants who had not received any dose of the vaccine).

Data analysis: a total of 14,161 children and adolescents aged 12–17 years who were tested for COVID-19 in the date range of the study, vaccinated or unvaccinated, were included in the study. The case and control groups were described using frequency distribution. The vaccine effectiveness of the BNT162b2 Pfizer vaccine among children/adolescents aged 12–17 years at least 14 days post-second dose was estimated by calculating relative risk reduction (RRR). The differences in VE of the COVID-19 vaccine according to age, gender, nationality (Qatari and non-Qatari), and vaccination status (fully vaccinated, partially vaccinated, unvaccinated) were also analyzed. VE based on number of days from receipt of second dose was also analyzed, because people who were vaccinated earlier are at increased risk for infection compared to those vaccinated later.

3. Results

Table 1 shows the characterization of the 14,161 children and adolescents aged 12–17 years included in the study by age, gender, nationalities, and vaccination status. The majority (40.6%) were 12–13 years old. The male to female proportions were nearly the same. A majority (60.9%) of the study participants were non-Qataris, as expected from the population distribution of Qatar. A total of 7925 (55.96%) were vaccinated with two doses of the BNT162b2 vaccine, and 6225 (43.96%) were unvaccinated. This higher proportion of vaccinated participants can be explained by the fact that as of 31 May 2021, more than half of residents had received at least one dose and 41% had completed both doses.

Table 1. Description of the study participants by age, gender, nationalities, and vaccine status.

	Fully Vaccinated (Two Doses)	Partially Vaccinated (One Dose)	Unvaccinated	Total	*p*-Value
		Distribution by Age group (in years)			
12–13	1962 (34.1%)	2 (0.03%)	3787 (65.8%)	5751 (40.6%)	<0.001
14–15	3222 (67.7%)	6 (0.13%)	1534 (32.2%)	4762 (33.6%)	
16–17	2741 (75.1%)	3 (0.08%)	904 (24.8%)	3648 (25.8%)	
Total	7925 (55.96%)	11 (0.08%)	6225 (43.96%)	14,161 (100.0%)	
		Distribution by Gender			
Male	3840 (54.1%)	4 (0.06%)	3253 (45.8%)	7097 (50.1%)	2.753
Female	4085 (57.8%)	7 (0.1%)	2972 (42.1%)	7064 (49.9%)	
Total	7925 (55.96%)	11 (0.08%)	6225 (43.96%)	14,161 (100.0%)	
		Distribution by Nationalities			
Qataris	2914 (52.7%)	3 (0.05%)	2616 (47.3%)	5533 (39.1%)	1.202
Non-Qataris	5011 (58.1%)	8 (0.09%)	3609 (41.8%)	8628 (60.9%)	
Total	7925 (55.96%)	11 (0.08%)	6225 (43.96%)	14,161 (100.0%)	

A majority (65.8%) of the younger age group (12–13 years) were unvaccinated, while three-fourths (75.1%) of those aged 16–17 were fully vaccinated. Hence, Table 1 shows a significant (p-value < 0.001) association between age and completion of the primary series of vaccines; that is, as the age increases, the proportion of fully vaccinated children increases. (Table 1) This can be explained by the fact the older age group were enthusiastic to get vaccinated as this would give them the privilege to go out into malls and restaurants. There was no significant difference between the genders with regards to the vaccination status.

No significant difference in vaccination status was noted between the nationals and non-nationals. A similar proportion of nationals and non-nationals were vaccinated with at least one dose. This may be explained by the fact that the government provided COVID vaccines free of cost universally to both nationals and non-nationals.

According to the data presented in Table 2, only 3.1% of the study population were infected and, among the infected, 44.14% were aged 12–13, whereas only half this proportion (20.7%) were of the older age group (16–17 years). The likelihood of testing positive for COVID-19 decreases with increasing age and vaccination status. This is in sync with the vaccination rates among the different age groups.

Table 2. Distribution of cases and controls by age, gender, and nationalities.

	COVID-19-Positive (Cases)	COVID-19-Negative (Control)	Total	*p*-Value
Distribution by Age group (in Years)				
12–13	196 (44.1%)	5555 (40.5%)	5751	0.044
14–15	156 (35.1%)	4606 (33.6%)	4762	
16–17	92 (20.8%)	3556 (25.9%)	3648	
Total	444 (100%)	13,717 (100%)	14,161	
Distribution by Gender				
Male	224 (50.5%)	6873 (50.1%)	7097	0.886
Female	220 (49.5%)	6844 (49.9%)	7064	
Total	444 (3.1%)	13,717 (100%)	14,161	
Distribution by Nationalities				
Qataris	207 (46.6%)	5326 (38.8%)	5533	<0.001
Non-Qataris	237 (53.4%)	8391 (61.2%)	8628	
Total	444 (100%)	13,717 (100%)	14,161	

There was a significant (<0.001) difference between the cases and controls with regards to nationality. More than half (53.4%) of the cases were non-Qataris, in comparison to 46.6% Qataris. Similarly, a higher proportion of the control group were non-Qataris (61.2%). This can be explained by the population distribution of the residents of Qatar.

However, no significant difference was found between the cases and controls with regards to gender. The males to female proportion was similar in both case and control groups.

Table 3 shows that out of 7925 fully vaccinated subjects, only 97 (1.2%) tested positive for SARS-CoV-2 by RT-PCR, whereas 346 (5.6%) tested positive among the 6225 in the unvaccinated population. Among the case group, 21.8% were fully vaccinated, while 77.9% were unvaccinated. Similarly, among the control group, 57.2% had received at least one dose of COVID vaccine, while 42.8% were unvaccinated. There is a significant difference in the proportion of fully vaccinated between the cases and controls ($p < 0.001$).

Table 3. Vaccination status and COVID-19 test results of the study participants.

Vaccination Status	RT-PCR Positive (Cases)	RT-PCR Negative (Controls)	Total
Fully vaccinated	97 (21.8%)	7828 (57.1%)	7925 (55.96%)
Partially vaccinated	1 (0.3%)	10 (0.1%)	11 (0.08%)
Unvaccinated	346 (77.9%)	5879 (42.8%)	6225 (43.96%)
Total	444 (100%)	13,717 (100%)	14,161

Relative risk reduction (RRR) was calculated as:

$$\frac{\text{vaccinated among cases} \times \text{unvaccinated among controls}}{\text{vaccinated among controls} \times \text{unvaccinated among cases}} = 0.21$$

$$\text{Vaccine effectiveness (VE)} = 1 - \text{RRR} = 79.0\%$$

Figure 2 shows that, out of a total of 98 individuals who received the vaccine and tested SARS-CoV-2 positive, 2 (1.9%) tested positive within 14 days of the first dose, while 3 (2.8%) tested positive between 15 and 30 days after the first dose. Another 2.8% tested positive within 14 days of receiving the second dose. It is seen that nearly half of the fully vaccinated participants (48.1%) tested positive after 90 days following the second dose, while only 15.7% tested positive within 31–60 days of receiving the second dose, and 17.6% were infected between 61–90 days.

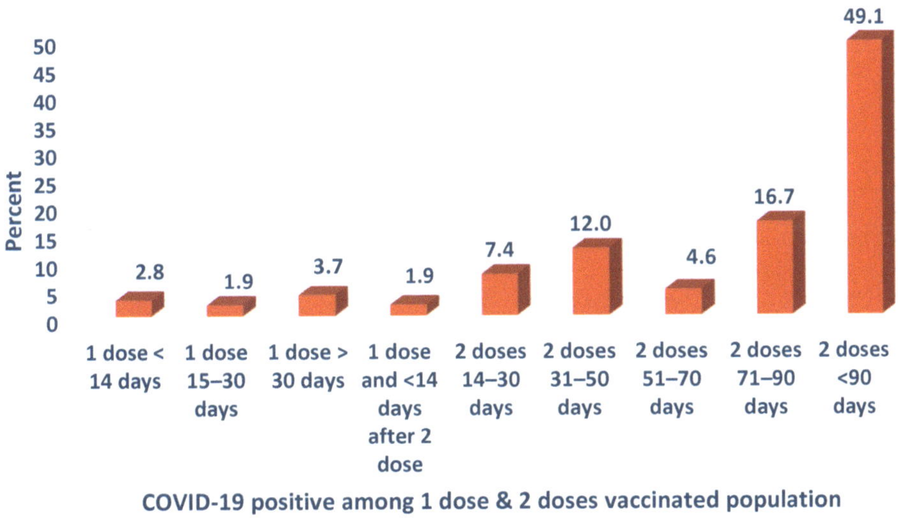

Figure 2. COVID-19 breakthrough infections among vaccinated study participants.

4. Discussion

The effectiveness against Alpha and Beta variants was high: over 75% in Qatar [8,28–30]. On the other hand, the effectiveness against Delta variant infection seven or more days after the second dose was (55.5%) (95% CI, 51.2–59.4%), irrespective of the vaccine type, 51.9% (95% CI, 47.0–56.4%) with BNT162b2, and 73.1% (95% CI, 67.5–77.8%) with mRNA-1273, specifically. The protection was higher 14 days after the second dose of the primary series. Delta variant's effectiveness was evaluated several months after the second dose for the residents [31].

The present study estimated the effectiveness of BNT162b2 (Pfizer) vaccine against SARS-CoV-2 infection, in children/adolescents aged 12–17 years, 14 or more days after the second dose, to be 79% in Qatar during the pre-Omicron period. Protection offered by the vaccine against SARS-CoV-2 infection among vaccinated children was higher among the older age group (p-value < 0.001). The VE showed a gradual decline in immunity over time following the second dose.

Another study carried out in Qatar during the pre-Omicron period found that the vaccine effectiveness against SARS-CoV-2 infection among adolescents was 87.6% (95% CI, 84.0 to 90.4). The level of protection was approximately 95% post-second dose and declined slowly over time but remained above 50% for at least five months [32].

The findings are consistent with evidence from other countries regarding protection provided by the vaccine in preventing SARS-CoV-2 infection among children and adolescents [33–38]. The vaccine demonstrated efficacy similar to that observed in young adults [39].

Testing for SARS-CoV-2 is performed on a mass scale in Qatar [40,41]. About three-fourths of cases are diagnosed because of routine screening tests and not because of the appearance of symptoms [42]. Since the hospitalization and deaths were low among the younger population, it was difficult to differentiate whether the protection was offered by the natural infection or by vaccination with the mRNA vaccines [40–44].

In a study carried out in Italy, the fully vaccinated group had a vaccine effectiveness of 29.4% (95% CI 28.5–30.2) and 41.1% (95% CI 22.2–55.4) with BNT162b2 (Pfizer-BioNTech) against SARS-CoV-2 infection and severe SARS-CoV-2 infection, respectively. The partially vaccinated group had a vaccine effectiveness of 27.4% (95% CI 26.4–28.4) against SARS-CoV-2 infection and 38.1% (95% CI 20.9–51.5) against severe SARS-CoV-2 infection. The vaccine effectiveness was highest, at 38.7% (95% CI 37.7–39.7%), within the first 14 days after completing the primary series. However, it decreased to 21.2% (95% CI 19.7–22.7%) between 43 and 84 days after being fully vaccinated with two doses [33].

In a retrospective cohort study, from Singapore, the estimated vaccine effectiveness (VE) against all COVID-19 infections in the age group of 12–18 years following two doses of BNT162b2 (Pfizer-BioNTech, New York, NY, USA) vaccines was 59% (95% CI: 55–63%) over the period of Delta variant dominance from 1 June to 20 November 2021 [39]. In a US study, during the Omicron-dominant period, the vaccine effectiveness was 59.0% (95% CI 22.0–79.0) 14–149 days after receipt of the second dose among adolescents aged 12–15 years [45].

The VISION Network study carried out across 10 states of the United States during 26 August 2021–22 January 2022 found that vaccine effectiveness after receipt of both two and three doses was lower during the Omicron-predominant period when compared to the Delta-predominant period. During both periods, VE waned with increasing time since vaccination. During the Omicron period, VE during the first two months after a third dose was 87% against emergency department/OPD clinic visits, and the VE decreased to 66% among those vaccinated who were vaccinated four to five months earlier. VE against hospitalizations was 91% during the first two months following a third dose and decreased to 78% beyond four months after a third dose [46].

A study conducted in England showed that vaccine effectiveness was 76.3% (95% CI 61.1–85.6%) 28 days after the first dose for those aged 16–17 years and 83.4% (54.0–94.0%) for those aged 12–15 years. The first dose of the vaccine was most effective for 16–17-year-olds against symptomatic disease caused by the Delta variant between days 14 and 20, with a peak effectiveness of 75.9% (95% CI: 74.3–77.3). However, effectiveness gradually decreased to 29.3% (25.9–32.6) between days 84 and 104. Among children/adolescents aged 12–15 years and 16–17 years, the VE against Delta infection showed a peak of 68% (95% CI: 64–71%) and 62% (95% CI: 57–66%), respectively, on days 21–48 after the first dose. Among those aged 16–17 years who received both doses, the VE against infection with the Delta variant was highest, at 93% (95% CI: 90–95%), between days 35 and 62 after vaccination but declined to 84% (95% CI: 76–89%) after 63 days [47].

Out of the 991,682 children and adolescents in Denmark who underwent RT-PCR testing for SARS-CoV-2, 7.5% (74,611) tested positive. Compared to unvaccinated adolescents, those who received one dose of the vaccine had an estimated effectiveness of 62% (with a 95% confidence interval of 59% to 65%) after 20 days. After 60 days, the estimated effectiveness of two doses was 93% (with a 92% to 94% confidence interval) during a period when the Delta variant was the most prevalent. The BNT162b2 vaccine demonstrated high effectiveness, of 93%, against SARS-CoV-2 infection among adolescents aged 12–17 years 60 days after receiving the second dose [48].

5. Strengths

Testing for SARS-CoV-2 was carried out on a mass scale in Qatar, and the results are tracked centrally. Nearly all testing in the country was via RT-PCR during this period. Universal access to the vaccines was provided to the eligible population free of cost. The National Vaccine registry facilitated obtaining the vaccination details of any individual vaccinated in the country. Additionally, since most of the testing was routine, contacts were likely found to be asymptomatic. These facts would suggest that the vaccine was efficacious against infection and not symptomatic infection/hospitalization/those seeking health care.

6. Limitations

Nearly all testing in the country was via RT-PCR during this period; however, there may have been a minority group who would have gone for home rapid testing kits. Since our study design relies on people being tested via RT-PCR only, these home tests were not included; moreover, a home test did not result in the same contract tracing as the PCR; hence, we may be underestimating cases. However, since nearly all testing in the country is performed using RT-PCR, these will not constitute huge numbers. Confounders like ethnicity and the presence of comorbidity have not been taken into consideration in this study. This study does not include the effectiveness of additional doses of the COVID-19 vaccine in severely immunocompromised children and adolescents, for whom additional doses should be considered as part of the primary vaccination schedule.

7. Conclusions

The BNT162b2 vaccine was associated with high protection against SARS-CoV-2 disease in children and adolescents. At this stage, priority should be given to completing the primary vaccination course for all the eligible population. Taking into consideration the waning of immunity, attention should be given to providing additional doses to high-risk and priority groups, according to national recommendations.

Unvaccinated persons are more likely to be infected when compared to vaccinated individuals, leading to higher incidence among the unvaccinated. Thus, Qatar began with easing restrictions for those vaccinated; however, now there is concern about a potentially increased risk of exposure among vaccinated individuals. Due to their perceived lower risk, the vaccinated may have adhered less strictly to safety measures such as face masks.

Author Contributions: Concept and designing the study, K.O., J.M., R.B., H.K. and S.A.; data collection, K.O., J.M., J.C., R.R., R.B., J.A., B.R., R.Y.E., K.B., A.M.E. and Z.A.T.; data collection coordination, K.O.; data cleaning and analysis, J.C., R.R. and J.A.; manuscript writing, K.O., J.M., J.C. and R.R.; manuscript review, J.M., J.C., R.R., R.B., H.K. and S.A.; guarantor, J.M. and S.A. All authors have read and agreed to the published version of the manuscript.

Funding: No separate funds were required for the study as it was based on secondary data. The dataset of this study is held at the Ministry of Public Health, Qatar.

Institutional Review Board Statement: Ethical approval was obtained from the Health Research Governance Department at the Ministry of Public Health. The Health Research Governance Department at the Ministry of Public Health waived informed consent. The researchers accessed data through a restricted-access agreement that prevents their sharing with a third party or publicly. All data were deidentified before sharing for analysis.

Informed Consent Statement: The Health Research Governance Department at the Ministry of Public Health waived informed consent because the study was based on secondary data and the data was already owned by MOPH and no contact with the participant population or no interference in the treatment they receive as the data is historical.

Data Availability Statement: The researchers accessed data through a restricted-access agreement that prevents their sharing with a third party or publicly.

Conflicts of Interest: The authors declare no conflict of interest.

References

1. Ng, Y.; Li, Z.; Chua, Y.X.; Chaw, W.L.; Zhao, Z.; Er, B.; Pung, R.; Chiew, C.J.; Lye, D.C.; Heng, D.; et al. Evaluation of the effectiveness of surveillance and containment measures for the first 100 patients with COVID-19 in Singapore—2 January–29 February 2020. *Morb. Mortal. Wkly. Rep.* **2020**, *69*, 307–311. [CrossRef] [PubMed]
2. Ministry of Public Health of Qatar. National COVID-19 Vaccination Program Data. COVID-19 National COVID-19 Vaccination Program Data. Available online: moph.gov.qa (accessed on 1 December 2022).
3. World Health Organization. 29 September 2021. Available online: https://www.who.int/publications/i/item/WHO-2019-nCoV-Sci_Brief-Children_and_adolescents-2021.1 (accessed on 30 November 2022).

4. Voysey, M.; Clemens, S.A.C.; Madhi, S.A.; Weckx, L.Y.; Folegatti, P.M.; Aley, P.K.; Angus, B.; Baillie, V.L.; Barnabas, S.L.; Bhorat, Q.E.; et al. Safety and effectiveness of the ChAdOx1 nCoV-19 vaccine (AZD1222) against SARS– CoV-2: An interim analysis of four randomised controlled trials in Brazil, South Africa, and the UK. *Lancet* **2021**, *397*, 99–111. [CrossRef]
5. Skowronski, D.M.; De Serres, G. Safety and Effectiveness of the BNT162b2 mRNA COVID-19 Vaccine. *N. Engl. J. Med.* **2021**, *384*, 1576–1577. [PubMed]
6. Baden, L.R.; El Sahly, H.M.; Essink, B.; Kotloff, K.; Frey, S.; Novak, R.; Diemert, D.; Spector, S.A.; Rouphael, N.; Creech, C.B.; et al. Effectiveness and Safety of the mRNA-1273 SARS-CoV-2 Vaccine. *N. Engl. J. Med.* **2021**, *384*, 403–416. [CrossRef] [PubMed]
7. European Centre for Disease Prevention and Control (ECDC). Overview of the Implementation of COVID-19 Vaccination Strategies and Vaccine Deployment Plans in the EU/EEA. Stockholm: ECDC. Available online: https://www.ecdc.europa.eu/en/publications-data/overview-implementation-covid-19-vaccinationstrategies-and-deployment-plans (accessed on 31 January 2022).
8. Chantal, V.; Fauver, J.; Grubaugh, N.D. Multiplexed RT-qPCR to Screen for SARS-CoV-2 B.1.1.7, B.1.351, and P.1 Variants of Concern V.3. 2021. Available online: https://www.protocols.io/view/multiplexed-rt-qpcr-to-screen-for-sars-cov-2-b-1-1-br9vm966 (accessed on 10 February 2021).
9. Abu-Raddad, L.J.; Chemaitelly, H.; Butt, A.A. National Study Group for Covid Vaccination. Effectiveness of the BNT162b2 COVID-19 Vaccine against the B.1.1.7 and B.1.351 Variants. *N. Engl. J. Med.* **2021**, *385*, 187–189. [CrossRef]
10. Butt, A.A.; Dargham, S.R.; Tang, P.; Chemaitelly, H.; Hasan, M.R.; Coyle, P.V.; Kaleeckal, A.H.; Latif, A.N.; Loka, S.; Shaik, R.M.; et al. COVID-19 disease severity in persons infected with the Omicron variant compared with the Delta variant in Qatar. *J. Glob. Health* **2022**, *12*, 05032. [CrossRef]
11. Chemaitelly, H.; Yassine, H.M.; Benslimane, F.M.; Al Khatib, H.A.; Tang, P.; Hasan, M.R.; Malek, J.A.; Coyle, P.; Ayoub, H.H.; Al Kanaani, Z.; et al. mRNA-1273 COVID-19 vaccine effectiveness against the B.1.1.7 and B.1.351 variants and severe COVID-19 disease in Qatar. *Nat. Med.* **2021**, *27*, 1614–1621. [CrossRef]
12. Qatar Viral Genome Sequencing Data: Data on Randomly Collected Samples. GISAID. 2021. Available online: https://www.gisaid.org/phylodynamics/global/nextstrain/ (accessed on 1 December 2022).
13. Benslimane, F.M.; Al Khatib, H.A.; Al Jamal, O.; Albatesh, D.; Boughattas, S.; Ahmed, A.A.; Bensaad, M.; Younuskunju, S.; Mohamoud, Y.A.; Badr, M.A.; et al. One Year of SARS-CoV-2: Genomic Characterization of COVID-19 outbreak in Qatar. 20 May 2021. Available online: https://www.medrxiv.org/content/10.1101/2021.05.19.21257433v2 (accessed on 1 December 2022).
14. Hasan, M.R.; Kalikiri, M.K.R.; Mirza, F.; Sundararaju, S.; Sharma, A.; Xaba, T.; Lorenz, S.; Chemaitelly, H.; El-kahlout, R.A.; Tsui, K.M.; et al. Real-time SARS-CoV-2 genotyping by high-throughput multiplex PCR reveals the epidemiology of the variants of concern in Qatar. *Int. J. Infect. Dis.* **2021**, *112*, 52–54. [CrossRef]
15. Russell, F.M.; Greenwood, B. Who should be prioritised for COVID-19 vaccination? *Hum. Vaccines Immunother.* **2021**, *17*, 1317–1321. [CrossRef]
16. Townsend, E. Debate: The impact of school closures and lockdown on mental health in young people. *Child Adolesc. Ment. Health* **2020**, *25*, 265–266. [CrossRef]
17. Jones, E.A.K.; Mitra, A.K.; Bhuiyan, A.R. Impact of COVID-19 on mental health in adolescents: A systematic review. *Int. J. Environ. Res. Public Health* **2021**, *18*, 2470. [CrossRef] [PubMed]
18. Frenck, R.W., Jr.; Klein, N.P.; Kitchin, N.; Gurtma, A.; Absalon, J.; Lockhart, S.; Perez, J.L.; Walter, E.B.; Senders, S.; Bailey, R.; et al. Safety, immunogenicity, and efficacy of the BNT162b2 COVID-19 vaccine in adolescents. *N. Engl. J. Med.* **2021**, *385*, 239–250. [CrossRef] [PubMed]
19. Abu-Raddad, L.J.; Chemaitelly, H.; Yassine, H.M.; Benslimane, F.M.; Khatib, H.A.A.; Tang, P.; Malek, J.A.; Coyle, P.; Ayoub, H.H.; Kanaani, Z.A.; et al. Pfizer-BioNTech mRNA BNT162b2 COVID-19 vaccine protection against variants of concern after one versus two doses. *J. Travel Med.* **2021**, *28*, taab083. [CrossRef]
20. Powell, A.A.; Kirsebom, F.; Stowe, J.; McOwat, K.; Saliba, V.; Ramsay, M.E.; Lopez-Bernal, J.; Andrews, N.; Ladhani, S.N. Effectiveness of BNT162b2 against COVID-19 in adolescents. *Lancet Infect. Dis.* **2022**, *22*, 581–583. [CrossRef] [PubMed]
21. Hanson, K.E.; Caliendo, A.M.; Arias, C.A.; Hayden, M.K.; Englund, J.A.; Lee, M.J.; Loeb, M.; Patel, R.; Alayli, A.E.; Altayar, O.; et al. IDSA Guidelines on the Diagnosis of COVID-19: Molecular Diagnostic Testing. Infectious Diseases Society of America. *Clin. Infect. Dis.* **2021**, ciab048. Available online: https://www.idsociety.org/practice-guideline/covid-19-guideline-diagnostics/ (accessed on 1 December 2022). [CrossRef]
22. Abu-Raddad, L.J.; Chemaitelly, H.; Ayoub, H.H.; Kanaani, Z.A.; Khal, A.A.; Kuwari, E.A.; Butt, A.A.; Coyle, P.; Jeremijenko, A.; Kaleeckal, A.H.; et al. Characterizing the Qatar advanced-phase SARS-CoV-2 epidemic. *Sci. Rep.* **2021**, *11*, 6233. [CrossRef] [PubMed]
23. Jeremijenko, A.; Chemaitelly, H.; Ayoub, H.H.; Alishaq, M.; Abou-Samra, A.-B.; Ajmi, J.A.A.A.A.A.; Ansari, N.A.A.A.; Kanaani, Z.A.; Khal, A.A.; Kuwari, E.A.; et al. Herd immunity against severe acute respiratory syndrome coronavirus 2 infection in 10 communities, Qatar. *Emerg. Infect. Dis.* **2021**, *27*, 1343–1352. [CrossRef]
24. Al-Tani, M.H.; Farag, E.; Bertollini, R.; Romaihi, H.E.A.; Abdeen, S.; Abdelkarim, A.; Daraan, F.; Ismail, A.I.H.E.; Mostafa, N.; Sahal, M.; et al. SARS-CoV-2 infection is at herd immunity in the majority segment of the population of Qatar. *Open Forum Infect. Dis.* **2021**, *8*, ab221. [CrossRef]

25. Coyle, P.V.; Chemaitelly, H.; Kacem, M.A.B.H.; Molawi, N.H.A.A.; Kahlout, R.A.E.; Gilliani, I.; Younes, N.; Anssari, G.A.A.A.A.; Kanaani, Z.A.; Khal, A.A.; et al. SARS-CoV-2 seroprevalence in the urban population of Qatar: An analysis of antibody testing on a sample of 112,941 individuals. *Iscience* **2021**, *24*, 102646. [CrossRef]
26. Butt, A.A.; Dargham, S.R.; Chemaitelly, H.; Al Khal, A.; Tang, P.; Hasan, M.R.; Coyle, P.V.; Thomas, A.G.; Borham, A.M.; Concepcion, E.G.; et al. Severity of Illness in Persons Infected With the SARS-CoV-2 Delta Variant vs. Beta Variant in Qatar. *JAMA Intern. Med.* **2022**, *182*, 197. [CrossRef]
27. Bertollini, R.; Chemaitelly, H.; Yassine, H.M.; Al-Thani, M.H.; Al-Khal, A.; Abu-Raddad, L.J. Associations of vaccination and of prior infection with positive PCR test results for SARS-CoV-2 in airline passengers arriving in Qatar. *JAMA* **2021**, *326*, 185–188. [CrossRef] [PubMed]
28. Abu-Raddad, L.J.; Chemaitelly, H.; Ayoub, H.H.; AlMukdad, S.; Yassine, H.M.; Al-Khatib, H.A.; Smatti, M.K.; Tang, P.; Hasan, M.R.; Coyle, P.; et al. Effect of MRNA Vaccine Boosters against SARS-CoV-2 Omicron Infection in Qatar. *N. Engl. J. Med.* **2022**, *386*, 1804–1816. [CrossRef] [PubMed]
29. Jackson, M.L.; Nelson, J.C. Test-negative design for estimating influenza vaccine effectiveness. *Vaccine* **2013**, *31*, 2165–2168. [CrossRef] [PubMed]
30. Verani, J.R.; Baqui, A.H.; Broome, C.V.; Cherian, T.; Cohen, C.; Farrar, J.L.; Feikin, D.R.; Groome, M.J.; Hajjeh, R.A.; Johnson, H.L.; et al. Case–control vaccine effectiveness studies: Preparation, design, and enrollment of cases and controls. *Vaccine* **2017**, *35*, 3295–3302. [CrossRef]
31. Tang, P.; Hasan, M.R.; Chemaitelly, H.; Yassine, H.M.; Benslimane, F.M.; Khatib, A.A.; AlMukdad, S.; Coyle, P.; Ayoub, H.H.; Kanaani, Z.A.; et al. BNT162b2 and mRNA-1273 COVID-19 vaccine effectiveness against the SARS-CoV-2 Delta variant in Qatar. *Nat. Med.* **2021**, *27*, 2136–2143. [CrossRef]
32. World Health Organization (WHO). WHO SAGE Roadmap for Prioritizing Uses of COVID-19 Vaccines. *Geneva*. 2022. Available online: https://www.who.int/publications/i/item/who-sage-roadmap-for-prioritizing-uses-ofcovid-19-vaccines (accessed on 28 June 2023).
33. Chemaitelly, H.; AlMukdad, S.; Ayoub, H.H.; Altarawneh, H.N.; Coyle, P.; Tang, P.; Yassine, H.M.; Al-Khatib, H.A.; Smatti, M.K.; Hasan, M.R.; et al. COVID-19 Vaccine Protection among Children and Adolescents in Qatar. *N. Engl. J. Med.* **2022**, *387*, 1865–1876. [CrossRef]
34. Sacco, C.; Del Manso, M.; Mateo-Urdiales, A.; Rota, M.C.; Petrone, D.; Riccardo, F.; Bella, A.; Siddu, A.; Battilomo, S.; Proietti, V.; et al. Effectiveness of BNT162b2 vaccine against SARSCoV-2 infection and severe COVID-19 in children aged 5–11 years in Italy: A retrospective analysis of January–April 2022. *Lancet* **2022**, *400*, 97–103. [CrossRef]
35. Levy, M.; Recher, M.; Hubert, H.; Javouhey, E.; Flechelles, O.; Leteurtre, S.; Angoulvant, F. Multisystem Inflammatory Syndrome in Children by COVID-19 Vaccination Status of Adolescents in France. *JAMA* **2022**, *327*, 281–283. [CrossRef]
36. Walter, E.B.; Talaat, K.R.; Sabharwal, C.; Gurtman, A.; Lockhart, S.; Paulsen, G.C.; Barnett, E.D.; Muñoz, F.M.; Maldonado, Y.; Pahud, B.A.; et al. Evaluation of the BNT162b2 COVID-19 Vaccine in Children 5 to 11 Years of Age. *N. Engl. J. Med.* **2022**, *386*, 35–46. [CrossRef]
37. Fowlkes, A.L.; Yoon, S.K.; Lutrick, K.; Gwynn, L.; Burns, J.; Grant, L.; Phillips, A.L.; Ellingson, K.; Farraris, M.V.; LeClair, L.B.; et al. Effectiveness of 2-Dose BNT162b2 (Pfizer BioNTech) mRNA Vaccine in Preventing SARS-CoV-2 Infection Among Children Aged 5–11 Years and Adolescents Aged 12-15 Years—PROTECT Cohort, July 2021–February 2022. *Morb. Mortal. Wkly. Rep.* **2022**, *71*, 422–428. [CrossRef]
38. Cohen-Stavi, C.J.; Magen, O.; Barda, N.; Yaron, S.; Peretz, A.; Netzer, D.; Giaquinto, C.; Judd, A.; Leibovici, L.; Hernán, M.A.; et al. BNT162b2 Vaccine Effectiveness against Omicron in Children 5 to 11 Years of Age. *N. Engl. J. Med.* **2022**, *387*, 227–236. [CrossRef] [PubMed]
39. Veneti, L.; Berild, J.D.; Watle, S.V.; Starrfelt, J.; Greve-Isdahl, M.; Langlete, P.; Bøås, H.; Bragstad, K.; Hungnes, O.; Meijerink, H. Vaccine effectiveness with BNT162b2 (Comirnaty, Pfizer BioNTech) vaccine against reported SARS-CoV-2 Delta and Omicron infection among adolescents, Norway, August 2021 to January 2022. *medRxiv* **2022**, 22272854. [CrossRef]
40. Ali, K.; Berman, G.; Zhou, H.; Deng, W.; Faughnan, V.; Coronado-Voges, M.; Ding, B.; Dooley, J.; Girard, B.; Hillebrand, W.; et al. Evaluation of mRNA-1273 SARS-CoV-2 Vaccine in Adolescents. *N. Engl. J. Med.* **2021**, *385*, 2241–2251. [CrossRef] [PubMed]
41. Chemaitelly, H.; Tang, P.; Hasan, M.R.; AlMukdadm, S.; Yassine, H.M.; Benslimane, F.M.; Khatib, H.A.A.; Coyle, P.; Ayoub, H.H.; Kanaani, Z.A.; et al. Waning of BNT162b2 vaccine protection against SARS-CoV-2 infection in Qatar. *N. Engl. J. Med.* **2021**, *385*, e83. [CrossRef]
42. Altarawneh, H.N.; Chemaitelly, H.; Ayoub, H.H.; Tang, P.; Hasan, M.R.; Yassibes, H.M.; Al-Khatib, H.A.; Smatti, M.K.; Coyle, P.; Al-Kananni, Z.; et al. Effects of previous infection and vaccination on symptomatic omicron infections. *N. Engl. J. Med.* **2022**, *387*, 21–34. [CrossRef]
43. Abu-Raddad, L.J.; Chemaitelly, H.; Bertollini, R. Severity of SARS-CoV-2 reinfections as compared with primary infections. *N. Engl. J. Med.* **2021**, *385*, 2487–2489. [CrossRef]
44. Altarawneh, H.N.; Chemaitelly, H.; Hasan, M.R.; Ayoub, H.H.; Qassim, S.; AlMukdad, S.; Coyle, P.; Yassine, H.M.; Al-Khatib, H.A.; Benslimane, F.M.; et al. Protection against the omicron variant from previous SARS-CoV-2 infection. *N. Engl. J. Med.* **2022**, *386*, 1288–1290. [CrossRef]

45. Chemaitelly, H.; Ayoub, H.H.; AlMukdad, S.; Coyle, P.; Tang, P.; Yassine, H.M.; Al-Khatib, H.A.; Smatti, M.K.; Hasan, M.R.; Al-Kanaani, Z.; et al. Duration of mRNA vaccine protection against SARS-CoV-2 omicron BA.1 and BA.2 subvariants in Qatar. *Nat. Commun.* **2022**, *13*, 3082. [CrossRef]
46. Ferdinands, J.M.; Rao, S.; Dixon, B.E.; Mitchell, P.K.; DeSilva, M.B.; Irving, S.A.; Lewis, N.; Natarajan, K.; Stenehjem, E.; Grannis, S.J.; et al. Waning 2-Dose and 3-Dose Effectiveness of mRNA Vaccines Against COVID-19-Associated Emergency Department and Urgent Care Encounters and Hospitalizations Among Adults during Periods of Delta and Omicron Variant Predominance—VISION Network, 10 States, August 2021–January 2022. *Morb. Mortal. Wkly. Rep.* **2022**, *71*, 255–263. [CrossRef]
47. Kildegaard, H.; Lund, L.C.; Hojlund, M.; Stensballe, L.G.; Pottegard, A. Risk of adverse events after covid-19 in Danish children and adolescents and effectiveness of BNT162b2 in adolescents: Cohort study. *BMJ* **2022**, *377*, 068898. [CrossRef]
48. Usherwood, T.; LaJoie, Z.; Srivastava, V. A model and predictions for COVID-19 considering population behavior and vaccination. *Sci. Rep.* **2021**, *11*, 12051. [CrossRef] [PubMed]

vaccines

MDPI

Review

Safety, Efficacy, and Immunogenicity of Varying Types of COVID-19 Vaccines in Children Younger Than 18 Years: An Update of Systematic Review and Meta-Analysis

Yan Tian [1,2,3,4,5], Long Chen [1,2,3,4,5] and Yuan Shi [1,2,3,4,5,*]

1 Department of Neonatology, Children's Hospital of Chongqing Medical University, Chongqing 400014, China
2 National Clinical Research Center for Child Health and Disorders, Chongqing 400014, China
3 Ministry of Education Key Laboratory of Child Development and Disorders, Chongqing 400014, China
4 China International Science and Technology Cooperation Base of Child Development and Critical Disorders, Chongqing 400014, China
5 Chongqing Key Laboratory of Pediatrics, Chongqing 400014, China
* Correspondence: shiyuan@hospital.cqmu.edu.cn

Abstract: Vaccination is one of the most effective measures for children as the epidemic progresses. However, there is a significant research gap in the meta-analysis of the COVID-19 vaccines for children younger than 18 years. This study is a comprehensive review of different COVID-19 vaccines. Published articles were retrieved from PubMed, Embase, and the Cochrane Library. Twelve randomized controlled trials (RCTs) of COVID-19 vaccines were included in the review until 21 October 2022. Most local and systemic adverse reactions were predominantly mild to moderate in severity and disappeared quickly after different types of vaccines. The subunit vaccine had the highest safety. The significant risk was lower in the subunit vaccine group after the initial (RR 1.66, 95% CI 1.26–2.17, p = 0.0003) and booster vaccination (RR 1.40, 95% CI 1.02–1.92, p = 0.04). Younger children had a more outstanding safety profile in the mRNA and inactivated vaccine groups. The humoral immune response was proportional to the number of doses in the inactivated and the adenovirus vaccine groups, and the strength of immunogenicity was negatively correlated with age in the inactivated vaccine. The mRNA and the subunit vaccines provided satisfactory prevention against COVID-19, especially seven days after the booster dose. However, more research and longer-term follow-up are needed to assess the duration of immune responses, efficacy, and safety.

Keywords: COVID-19 vaccine; SARS-CoV-2; children; adverse reactions; immunogenicity; efficacy; meta-analysis

check for updates

Citation: Tian, Y.; Chen, L.; Shi, Y. Safety, Efficacy, and Immunogenicity of Varying Types of COVID-19 Vaccines in Children Younger Than 18 Years: An Update of Systematic Review and Meta-Analysis. *Vaccines* **2023**, *11*, 87. https://doi.org/10.3390/vaccines11010087

Academic Editor: Kay Choong See

Received: 30 November 2022
Revised: 24 December 2022
Accepted: 27 December 2022
Published: 30 December 2022

1. Introduction

Since the end of 2019, the novel coronavirus (COVID-19) has become a public health threat to people [1], and the pandemic is having an unprecedented impact on the physical and mental health of people around the world [2]. Compared with adults, the proportion of COVID-19 cases in children and adolescents is lower. Although children with COVID-19 seem to have milder symptoms and may even be completely asymptomatic once infected [3–5], children can have severe diseases that result in hospitalization, and approximately one-third of adolescents hospitalized for COVID-19 were admitted to an intensive care unit, and 4.9% received invasive mechanical ventilation [6]. In addition, children infected with COVID-19 can develop serious complications, such as multisystem inflammatory syndrome (MIS-C), a severe but rare condition associated with COVID-19, which is a condition where different body parts become inflamed [7,8].

The consequences of the pandemic on children's development could be vast, with impacts likely on self-control, social competence, and other cognitive abilities [9]. Growing research informs the heavy psychosocial implications of the COVID-19 pandemic, bringing about mental health problems such as anxiety, depression, stress, and maladaptive behavior [10–13]. According to UNESCO's report, rising COVID-19 infection rates led to school closures around the world, but limiting the spread of COVID-19 through school closure may lead to reduce interaction with peers, lessen opportunities for physical exercise, and exacerbate adverse psychosocial health outcomes in children, and they have made little or no progress while learning from home [14–17]. An estimated 1800 schools have had school closures attributable to COVID-19 outbreaks, and more than 900,000 students have been affected [18]. In addition, it was estimated that approximately 1.5 billion young people worldwide had been forced to stay at home, which negatively influenced their social functioning [19].

Moreover, Omicron spreads more easily among children than the previous variants [20], and unvaccinated individuals provide opportunities for more variants to emerge [21]. Therefore, there is an urgent need to vaccinate children against COVID-19 to protect pediatric age groups from harm. A safe and effective vaccine is critically important for infants and young children. Vaccination is one of the effective measures to fight against COVID-19, which can help to reduce the rate of severe diseases [22]. However, there is insufficient evidence that receiving COVID-19 vaccines reduces child mortality or prevents the further spread of the disease, that younger children are at greater risk of spreading COVID-19, or that herd immunity can be achieved through it. Vaccinating children against COVID-19 can protect their mental health [23]. Vaccination reduces family damage due to parental illness, failing economies, and chronic stress [24]. Acquiring the COVID-19 vaccine could provide direct benefits to childhood education by allowing a safer return to school to secure their continued access to education, and letting parents return to full-time work to make the economy recover [25,26]. Therefore, there is an urgent need to protect children through vaccination.

Whether children and adolescents should be vaccinated against COVID-19 remains controversial. Children and adolescents are unique, and parents usually hesitate to vaccinate their children. The vaccine's novelty and safety concerns can hinder acceptance in the population [27,28]. Several studies and systematic reviews have been performed to demonstrate the safety, immunogenicity, and efficacy of the COVID-19 vaccine. However, there is a lack of experimental data to confirm the safety, efficacy, and immunogenicity of COVID-19 vaccines in children under three years of age and even in infants, as well as experimental data on the different types of vaccines in children younger than 18 years. Therefore, we aimed to comprehensively synthesize the evidence for the safety, efficacy, and immunogenicity of varying types of COVID-19 vaccines in children younger than 18 years as an update to these previously performed systematic reviews.

2. Materials and Methods

The systematic review and meta-analysis were conducted following the Preferred Reporting Items for Systematic Reviews and Meta-Analyses (PRISMA) guidelines, and the protocol was registered on PROSPERO (CRD42022369708).

2.1. Search Strategy

A systematic retrieval was performed on three databases (PubMed, Embase, and the Cochrane Library) from inception to 21 October 2022. The key search terms were as follows: infant, child, adolescent, COVID-19 vaccines, SARS-CoV-2, COVID-19, randomized controlled trial, and so on (The search details can be found in Tables S1–S3). The clinical trials registers (Clinical Trials.gov, an ongoing NIH trial registry) was also searched for related articles.

2.2. Selection Criteria

The inclusion of studies was based on the following criteria: (1) vaccines administered to children aged < 18 years; (2) RCTs comparing COVID-19 vaccines with other vaccines, placebo, adjuvant; (3) reported measures of safety (local and systemic adverse events), immunogenicity (noninferiority of geometric mean titers (GMTs)) or efficacy (COVID-19 infection). The exclusion criteria for the studies were as follows: (1) non-original studies; review, meta-analysis, systematic review, comments, letters, standards, guidelines, or conference abstracts; (2) non-RCT studies, including cohort studies, case-control studies, single-arm studies, cross-sectional studies; (3) animal models; and (4) outcomes without interest.

2.3. Data Extraction

After eliminating duplicates, two reviewers (Tian and Chen) screened titles and abstracts and then used predefined criteria to filter the full text of potentially relevant articles. Two authors independently extracted the following information from each of the included studies as outcome indicators: (1) name of the first author, date of publication, intervention measures (vaccine type, number of doses, adjuvant addition, and adjuvant type, etc.), sample size, intervention details; (2) the incidence of adverse events post-vaccination, including total adverse reactions, local adverse reactions, systemic adverse reactions and any specific adverse reactions, such as injection pain, erythema/redness, fever and so on; (3) humoral immune responses and cellular immune responses, including the seroconversion, geometric mean titers (GMT) after vaccination; and (4) incidence of confirmed COVID-19 after vaccination. In the case of differences, a consensus was reached through discussion or consultation with the third author (Shi). Immunogenicity was expressed through the noninferiority of the immune response. The noninferiority criterion indicated if the lower boundary of the 95% confidence interval for the geometric mean ratio was at least 0.67, with or without the difference in the percentage of participants with a serologic response was −10 percentage points or more. The seroconversion was defined as a geometric mean titer (GMT) increase of at least a fourfold increase from baseline after vaccination. The definition of COVID-19 was according to which participants were diagnosed with COVID-19 and if they were positive for SARS-CoV-2 by RT-PCR and with one or more associated symptoms. We carefully read the included studies' original text and Supplementary Materials to avoid missing data.

2.4. Data Analysis

All data were performed using RevMan 5.4.1 statistical software to pool dichotomous through its internal procedures, even if the number of events is 0 in the observation and/or control group. When I^2 values were > 50%, the random effects model was applied to pool the overall results; otherwise, the fixed effects model was used. This study used the risk ratio (RR), and 95% confidence interval (CI) in the case of dichotomous data (RR > 1 represented a risk effect). The I^2 statistic was used to assess the level of statistical heterogeneity. The RR was determined using the formula $RR = \frac{\text{incidence in exposed}}{\text{incidence in unexposed}} = \frac{a/(a+b)}{c/(c+d)}$ (details can be found in Table 1). The data were deemed heterogeneous when the I^2 values > 50%. p values less than 0.05 were considered, and this difference was statistically significant. If we detected heterogeneity, subgroup analyses were conducted to explore the source of heterogeneity. We performed subgroup analyses according to the number of vaccinations, type of vaccines, age of the recipients, and specific adverse reactions and considered sensitivity analyses by excluding pooled studies one by one. To appraise the methodological quality of the included studies, two reviewers (YT and LC) independently assessed each study's risk according to the Cochrane collaboration tool for assessing the risk of bias as high, low, or unclear for each item. The funnel plot and Egger test were used to judge the publication bias.

$$RR = \frac{\text{incidence in exposed}}{\text{incidence in unexposed}} = \frac{a/(a+b)}{c/(c+d)} \quad (1)$$

Formula (1) The formula for calculating RR.

Table 1. The four-cell table for calculating RR of RCTs.

	Develop Outcome	Do Not Develop Outcome
Exposed	a	b
Not Exposed	c	d

3. Results

3.1. Characteristics of Included Studies

As shown in the flow diagram in Figure 1, this study found 2276 research articles using the previously mentioned search terms. After removing duplicates, we screened 1505 records based on title and abstract, of which 1454 were determined to be irrelevant. Fifty-one articles were retrieved for full-text assessment. Finally, 12 articles were included in our analysis: 12 articles for safety, seven for immunogenicity, and five for the efficacy of COVID-19 vaccines. These 12 RCTs included four types of COVID-19 vaccines (mRNA, subunit, inactivated, and adenoviral vector vaccines). All included studies reported COVID-19 vaccines from eight countries and regions; a total of 17,731 participants that received the COVID-19 vaccine and 7444 participants who received a placebo were included in this study ranging in age from six months to 17 years old. Frenck et al. and Walter et al. did not provide the exact number of participants in the vaccine group and placebo group in the safety analysis, so we obtained the available data by calculating the product using the form of totals and percentages. Notably, two RCTs [29,30] received a total of three doses. The characteristics of the included studies are summarized in Table 2. We performed the quality assessment for those included studies, the methodological quality of the included studies was high, and the risk of bias was low. Incomplete data and other biases dominated those bias risks. In three studies, incomplete data due to a lack of reasons for exclusive participants during the experiment, two did not specify the method of concealment allocation, and one included a small number of participants, as shown in detail in Figures 2 and 3. Since there were slight differences in outcome indicators among the included studies, this analysis tested publication bias through seven RCTs. Publication bias was performed by funnel plot and Egger's test, and the results did not show evidence of publication bias in total, systemic, or local adverse reactions ($p < 0.05$) (Figures S1 and S2) but did show in the neutralizing antibody 28 days after dose 2 ($p < 0.05$) (Figure S3).

3.2. Safety of the COVID-19 Vaccines

3.2.1. Adverse Reactions to Different Introduction Doses

Results showed that the total, systemic, and local adverse reactions after vaccination, both in the mRNA and the adenovirus vector vaccine group, showed a significantly increased risk, and the risk was higher in the second dose than in the first dose (Figure S4 and Table 3). The same outcome was observed in the subunit vaccine in the total adverse reactions, but it should be noted that the risk of local adverse reactions was higher in the first dose (RR 2.93, 95%CI 1.76–4.89, $p < 0.0001$; Figure S4 and Table 3) than the second dose (RR 1.99, 95%CI 1.24–3.18, $p = 0.004$; Figure S4 and Table 3) in the subunit vaccine group. There was no difference in systemic adverse reactions of the subunit vaccine. Of note, we found that only the risk of local reactions after initial vaccination was statistically significant in the inactivated vaccine group (RR 6.34, 95%CI 1.54–26.10, $p = 0.01$; Figure S4 and Table 3). The heterogeneity among the above analyses was considerable, and we subsequently performed subgroup analysis for the specific adverse reactions in different vaccine groups.

In the mRNA vaccine groups, we found that the risk of most specific adverse reactions was higher after the booster dose, such as erythema or redness (RR 7.73, 95%CI 3.76–15.90, $p < 0.00001$; Figure S5, Table 4), swelling or hardness (RR 8.59, 95%CI 4.86–15.19, $p < 0.00001$; Figure S5, Table 4), fever (RR 7.85, 95%CI 2.58–23.91, $p = 0.0003$; Figure S5 and Table 4) and chills (RR 4.37, 95%CI 3.14–6.09, $p < 0.00001$; Figure S5 and Table 4). However, the risk of headache, arthralgia, nausea or vomiting, and loss of appetite after the initial vaccination and the risk of diarrhea and sleepiness after booster vaccination were of no significant difference in the mRNA vaccine group (Figure S5 and Table 4).

Table 2. The characteristics of the included studies.

Author, Year	Country	Phase	Age (Years)	Type of Vaccine	Dose of Administration (Per Dose)	Time of Inoculations (Days)	Control	No. of the Observation Group	No. of the Control Group
Ali et al. [31]	The USA	phase 2–3	12–17	mRNA-1273 vaccine (mRNA vaccine)	100 μg	0, 28	Saline	2486	1240
Anderson et al. [32]	the USA, Canada	phase 2–3	6 Months-5	mRNA-1273 vaccine (mRNA vaccine)	25 μg, 50 μg	0, 28	Saline	5011	1751
Áñez et al. [33]	the USA, Mexico	phase 3	12–17	NVX-CoV2373 (subunit vaccine)	0.5 mL	0, 21	Saline	1487	745
Buddy Creech et al. [34]	the USA, Canada	phase 2–3	6–11	mRNA-1273 vaccine (mRNA vaccine)	50 μg	0, 28	Saline	3385	995
Frenck et al. [35]	The USA	phase 3	12–15	BNT162b2 Vaccine (mRNA vaccine)	30 μg	0, 21	Saline	1131	1129
Han et al. [36]	China	phase 1–2	3–17	CoronaVac (Inactivated vaccine)	1.5 μg, 3.0 μg	0, 28	Aluminum hydroxide adjuvant	436	114
Liu et al. [37]	China	phase 2	12–17	MVC-COV1901 (subunit vaccine)	0.5 mL	0, 28	Saline	341	58
Thuluva et al. [38]	India	phase 2–3	5–17 (<12-≥5, <18-≥12)	CORBEVAX™ (subunit vaccine)	0.5 mL	0, 28	Placebo (Not noted)	468	156
Walter et al. [39]	the USA, Spain, Finland, Poland	phase 2–3	5–11	BNT162b2 Vaccine (mRNA vaccine)	10 μg	0, 21	Saline	1518	750
Xia et al. [29]	China	phase 1–2	3–17 (3–5, 6–12, 13–17)	WBIBP-CorV (Inactivated vaccine)	2.5 μg, 5 μg, 10 μg	0, 28, 56	Alum	612	204
Xia et al. [30]	China	phase 1–2	3–17 (3–5, 6–12, 13–17)	BBIBP-COV (Inactivated vaccine)	2 μg,4 μg, 8 μg	0, 28, 56	Saline and aluminum hydroxide adjuvant	756	252
Zhu et al. [40]	China	phase 2	6–17	Ad5-vectored COVID-19 vaccine (Adenovirus vaccine)	0.3 mL	0, 56	Placebo containing the same excipients as the vaccine, without viral particles	100	50

In addition, in the inactivated vaccine group, the data showed only the risk of local pain after initial vaccination (RR 21.53, 95%CI 3.00–154.35, $p = 0.002$; Figure S6 and Table S4) and booster vaccination (RR 6.84, 95%CI 1.96–23.90, $p = 0.003$; Figure S6 and Table S4) was significantly higher than in the control group, the risk of other specific adverse reactions was of no significant difference compared with the control group. Similar differences were observed in the subunit vaccine, only the risk of local pain after initial vaccination (RR 2.91, 95%CI 1.74–4.84, $p < 0.0001$; Figure S7 and Table S5) and booster vaccination (RR 1.97, 95%CI 1.23–3.16, $p = 0.005$; Figure S7 and Table S5) was statistically significant. Additionally, the data showed that only the risk of local pain (RR 5.67, 95%CI 1.83–17.55, $p = 0.003$; Figure S8 and Table S6) and fever (RR 7.00 95%CI 1.74–28.21, $p = 0.006$; Figure S8 and Table S6) after initial vaccination was statistically significant in the adenovirus vector vaccine.

After pooling whole available data on specific adverse reactions, the significant risk was higher in all vaccine groups than the control group but relatively lower in the subunit vaccine group, both after initial vaccination (RR 1.66, 95% CI 1.26–2.17, $p = 0.0003$; Table 5) and after booster vaccination (RR 1.40, 95% CI 1.02–1.92, $p = 0.04$; Table 5).

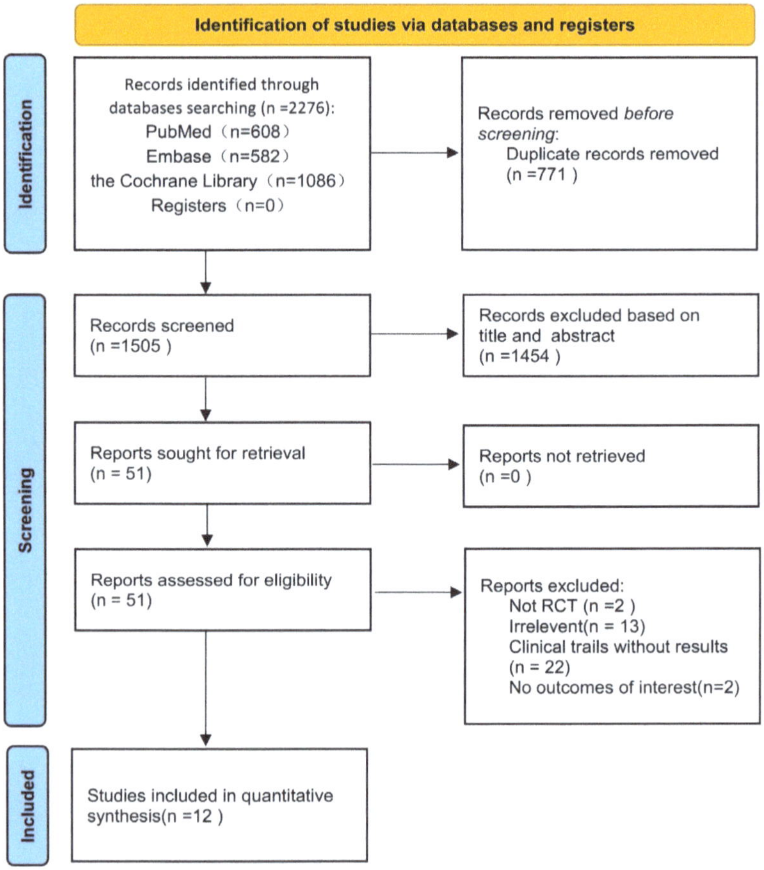

Figure 1. Flow chart of study identification and selection.

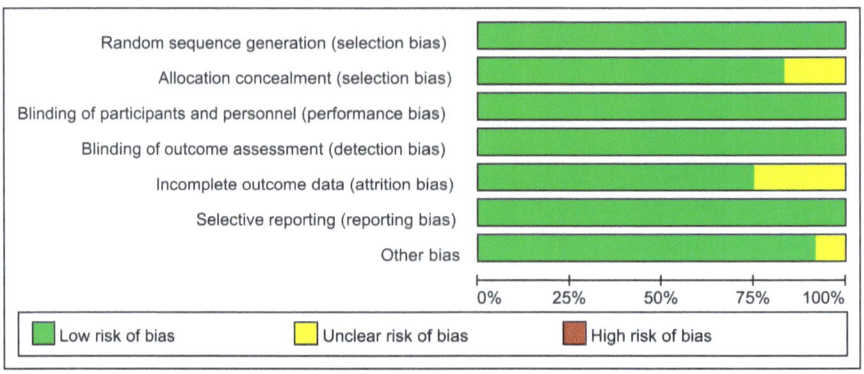

Figure 2. Risk of bias graph for included RCTs.

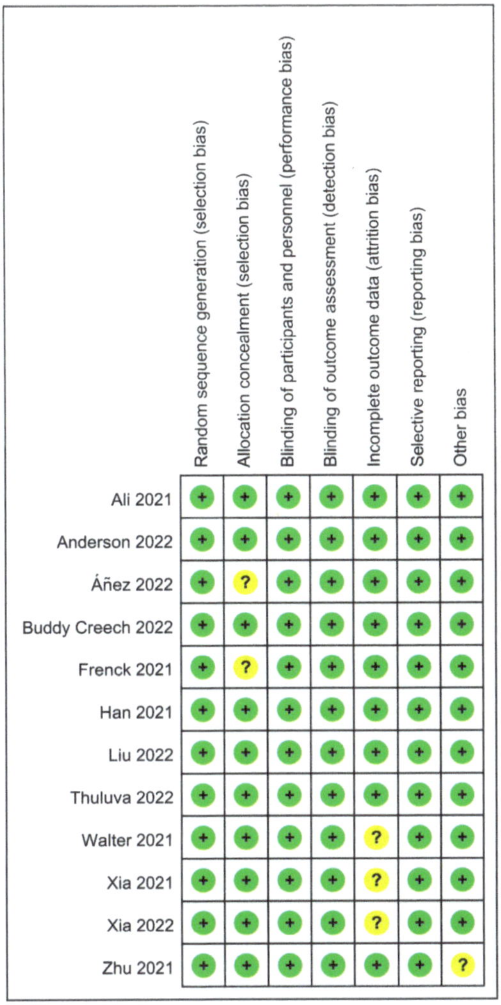

Figure 3. Risk of bias summary for included RCTs (the green color and special symbol "+" are represented a low risk of bias, and the yellow color and special symbol "?" are represented an unclear risk of bias).

3.2.2. Adverse Reactions to Different Age Groups

We observed high heterogeneity in the mRNA vaccine group when subgroup analysis was performed according to different vaccine types. However, the heterogeneity decreased without the RCT study by Anderson et al. The RCT by Anderson et al. was aimed at younger children aged six months to five years, the other four RCTs were conducted in children and adolescents above five years old. Therefore, we performed a subgroup analysis for specific adverse reactions of mRNA vaccine recipients of different ages. For children aged 12–17 years, the risk of almost specific adverse reactions after vaccination was significantly higher, especially erythema/redness (RR 10.74, 95%CI 2.72–43.37, $p = 0.0007$; Figure S9 and Table 6) and swelling/hardness (RR 10.61, 95%CI 4.13–27.28, $p < 0.00001$; Figure S9 and Table 6) after the first vaccination and erythema/redness (RR 10.16, 95%CI 2.05–50.29, $p = 0.005$; Figure S9 and Table 6), swelling/hardness (RR 10.00, 95%CI 2.11–47.24, $p = 0.004$; Figure S9 and Table 6) and fever (RR 15.28, 95%CI 10.11–23.11, $p < 0.00001$;

Figure S9 and Table 6) after the second vaccination. However, there were no statistical differences in headache (RR 1.35, 95%CI 1.00–1.82, $p = 0.05$; Figure S9 and Table 6) and nausea/vomiting (RR 1.78, 95%CI 0.82–3.86, $p = 0.14$; Figure S9 and Table 6) after the first vaccination. For younger children aged six months–11 years, the risk of swelling/hardness (RR 4.39, 95%CI 2.24–8.58, $p < 0.0001$; Figure S9 and Table 6) after the first vaccination and erythema/redness (RR 6.45, 95%CI 2.90–14.31, $p < 0.00001$; Figure S9 and Table 6), swelling/hardness (RR 7.71, 95%CI 4.33–13.72, $p < 0.00001$; Figure S9 and Table 6) after the booster vaccination were significantly higher. Subsequently, we compared various adverse reactions to vaccination occurrence in older and younger children following the mRNA vaccine. The data suggest a significantly higher risk of specific adverse responses in children aged 12–15 years versus 5–11 years after the booster vaccination (RR 1.84, 95%CI 1.25–2.72, $p = 0.002$; Figure S10). However, there was no statistical difference after the initial vaccination (RR 1.31, 95%CI 0.94–1.82, $p = 0.11$; Figure S10), indicating again that older children were at greater risk of adverse reactions after vaccination than younger children. Anderson et al. chose the mRNA-1273 vaccine as the intervention for children aged six months to five years, and we decided to directly compare the occurrence of various adverse reactions following mRNA-1273 vaccination in children aged 6–23 months and two to five years. Results showed that the risk of various adverse reactions in participants aged 6–23 months was significantly lower than two to five years both after the initial vaccination (RR 0.74, 95%CI 0.71–0.77, $p < 0.00001$; Figure S11 and Table 7) and the booster vaccination (RR 0.80, 95%CI 0.77–0.83, $p < 0.00001$; Figure S11 and Table 7). Overall, the risk of various adverse reactions after mRNA vaccination appears to be higher in older children aged 12–17 years than in younger children aged six months–11 years. A similar outcome was observed in children aged 6–23 months and two to five years, indicating again that younger children may have a greater safety profile in the mRNA vaccine.

Table 3. All adverse reactions in the vaccination group versus the control group.

		No. of Studies	RR (95% CI)	I^2	*p*-Value
Total Adverse Reactions					
mRNA vaccine	After dose 1	3	1.30 [1.07, 1.57]	98%	<0.05
	After dose 2	3	1.43 [1.14, 1.79]	98%	<0.05
Inactivated vaccine	After dose 1	1	1.27 [0.76, 2.13]	Not applicable	>0.05
	After dose 2	1	1.83 [0.90, 3.72]	Not applicable	>0.05
Submit vaccine	After dose 1	1	1.57 [1.17, 2.11]	Not applicable	<0.05
	After dose 2	1	1.94 [1.26, 2.98]	Not applicable	<0.05
Adenovirus vector vaccine	After dose 1	1	3.44 [1.78, 6.65]	Not applicable	<0.05
	After dose 2	1	8.25 [2.06, 33.00]	Not applicable	<0.05
Systemic adverse reactions					
mRNA vaccine	After dose 1	3	1.13 [1.02, 1.24]	88%	<0.05
	After dose 2	3	1.47 [1.08, 2.01]	99%	<0.05
Inactivated vaccine	After dose 1	1	1.32 [0.87, 2.00]	Not applicable	>0.05
	After dose 2	1	1.61 [0.76, 3.40]	Not applicable	>0.05
Submit vaccine	After dose 1	1	1.11 [0.78, 1.57]	Not applicable	>0.05
	After dose 2	1	1.22 [0.72, 2.09]	Not applicable	>0.05
Adenovirus vector vaccine	After dose 1	1	3.70 [1.55, 8.83]	Not applicable	<0.05
	After dose 2	1	6.00 [1.48, 24.38]	Not applicable	<0.05
Local adverse reactions					
mRNA vaccine	After dose 1	3	1.80 [1.11, 2.92]	99%	<0.05
	After dose 2	3	1.93 [1.25, 2.97]	99%	<0.05
Inactivated vaccine	After dose 1	1	6.34 [1.54, 26.10]	Not applicable	<0.05
	After dose 2	1	4.29 [1.03, 17.96]	Not applicable	=0.05
Submit vaccine	After dose 1	1	2.93 [1.76, 4.89]	Not applicable	<0.05
	After dose 2	1	1.99 [1.24, 3.18]	Not applicable	<0.05
Adenovirus vector vaccine	After dose 1	1	6.00 [1.94, 18.53]	Not applicable	<0.05
	After dose 2	1	19.69 [1.21,319.62]	Not applicable	<0.05

Note: RR, Risk ratio; CI, confidence interval; $p < 0.05$.

Table 4. Specific adverse reactions in the mRNA vaccine group versus the control group after dose 1 and dose 2.

		After Dose 1			After Dose 2		
	No. of Studies	RR (95% CI)	I^2	*p*-Value	RR (95% CI)	I^2	*p*-Value
Overall	5	1.91 [1.70, 2.16]	97	<0.05	3.13 [2.73, 3.59]	97	<0.05
Local pain	5	2.32 [1.72, 3.13]	98	<0.05	2.54 [1.89, 3.42]	98	<0.05
Erythema or Redness	5	5.66 [2.75, 11.65]	92	<0.05	7.73 [3.76, 15.90]	92	<0.05
Swelling or Hardness	5	6.21 [3.14, 12.28]	90	<0.05	8.59 [4.86, 15.19]	84	<0.05
Axillary Swelling	3	1.85 [1.15, 2.98]	93	<0.05	2.90 [2.02, 4.18]	84	<0.05
Fever	5	3.31 [1.47, 7.45]	92	<0.05	7.85 [2.58, 23.91]	96	<0.05
Headache	4	1.14 [0.92, 1.43]	94	>0.05	2.04 [1.63, 2.56]	94	<0.05
Fatigue	4	1.29 [1.16, 1.43]	79	<0.05	2.08 [1.70, 2.54]	93	<0.05
Myalgia	4	1.59 [1.39, 1.81]	43	<0.05	2.87 [2.07, 3.98]	90	<0.05
Arthralgia	4	1.10 [0.84, 1.45]	77	>0.05	2.22 [1.50, 3.28]	89	<0.05
Nausea or Vomiting	4	1.41 [0.99, 1.99]	75	=0.05	2.55 [2.23, 2.92]	0	<0.05
Chills	4	1.63 [1.15, 2.33]	89	<0.05	4.37 [3.14, 6.09]	86	<0.05
Diarrhea	2	1.27 [0.96, 1.67]	23	>0.05	1.21 [0.82, 1.80]	55	>0.05
Irritability or Crying	1	1.08 [1.01, 1.16]	Not applicable	<0.05	1.10 [1.01, 1.19]	Not applicable	<0.05
Sleepiness	1	0.97 [0.86, 1.09]	Not applicable	>0.05	1.04 [0.90, 1.19]	Not applicable	>0.05
Loss of appetite	1	1.12 [0.96, 1.30]	Not applicable	>0.05	1.25 [1.06, 1.48]	Not applicable	<0.05

Note: RR, Risk ratio; CI, confidence interval; $p < 0.05$.

Table 5. Overall specific adverse reactions among the vaccination group versus the control group after dose 1 and dose 2.

		After Dose 1			After Dose 2		
	No. of Studies	RR (95% CI)	I^2	*p*-Value	RR (95% CI)	I^2	*p*-Value
mRNA vaccine	5	1.91 [1.70, 2.16]	97	<0.05	3.13 [2.73, 3.59]	97	<0.05
Inactivated vaccine	2	1.76 [1.20, 2.57]	38	<0.05	2.18 [1.30, 3.67]	0	<0.05
Subunit vaccine	1	1.66 [1.26, 2.17]	19	<0.05	1.40 [1.02, 1.92]	12	<0.05
Vectored vaccine	1	5.27 [2.80, 9.91]	0	<0.05	6.21 [2.40, 16.11]	0	<0.05

Note: RR, Risk ratio; CI, confidence interval; $p < 0.05$.

One RCT [30] did not provide information on adverse reactions after the whole vaccination procedure of the inactivated vaccine in different age groups. So, a subgroup analysis was performed according to the age of the participants with two other RCTs. The data showed that only diarrhea (RR 0.21, 95%CI 0.05–0.93, $p < 0.05$; Figure S12 and Table S7) was statistically significant in children younger than 12 years old. In addition, the risk of overall specific adverse reactions was higher in recipients aged 12–17 years than in 3–12 years (RR 2.05, 95%CI 1.58–2.66, $p < 0.00001$; Figure S13), this was consistent with the subgroup analyses in mRNA vaccines, in which younger children may have a greater safety profile.

Subgroup analysis was conducted in the subunit vaccine, in children older than 12 years, only the risks of erythema/redness and nausea/vomiting were not statistically significant, while in children younger than 12 years, all adverse events were not statistically significant (Figure S14 and Table S8). Additionally, there was no significant difference in different age groups (RR 1.22, 95%CI 0.87–1.71, $p > 0.05$; Figure S15).

Further subgroup analysis could not be performed for the adenovirus vector vaccine due to insufficient data for the different age groups.

Table 6. Specific adverse reactions in mRNA vaccine recipients of different ages.

		≥12 Years				<12 Years			
		No. of Studies	RR (95% CI)	I^2	*p*-Value	No. of Studies	RR (95% CI)	I^2	*p*-Value
After dose 1	Local pain	2	3.15 [2.27, 4.37]	96	<0.05	3	1.89 [1.44, 2.48]	97	<0.05
	Erythema or Redness	2	10.74 [2.72, 43.37]	89	< 0.05	3	3.78 [1.96, 7.29]	89	<0.05
	Swelling or Hardness	2	10.61 [4.13, 27.28]	81	<0.05	3	4.39 [2.24, 8.58]	85	<0.05
	Fever	2	5.00 [1.40, 17.82]	89	<0.05	3	2.50 [1.02, 6.17]	90	=0.05
	Headache	2	1.35 [1.00, 1.82]	96	=0.05	2	0.98 [0.89, 1.07]	0	>0.05
	Fatigue	2	1.38 [1.24, 1.55]	71	<0.05	2	1.19 [1.03, 1.38]	72	<0.05
	Myalgia	2	1.70 [1.52, 1.90]	14	<0.05	2	1.40 [1.18, 1.67]	0	<0.05
	Arthralgia	2	1.33 [1.15, 1.55]	0	<0.05	2	0.84 [0.45, 1.56]	85	>0.05
	Nausea or Vomiting	2	1.78 [0.82, 3.86]	81	>0.05	2	1.24 [0.67, 2.27]	62	>0.05
	Chills	2	2.08 [1.31, 3.30]	92	<0.05	2	1.25 [0.83, 1.87]	69	>0.05
After dose 2	Local pain	2	3.64 [2.55, 5.19]	95	<0.05	3	1.99 [1.70, 2.34]	92	<0.05
	Erythema or Redness	2	10.16 [2.05, 50.29]	93	<0.05	3	6.45 [2.90, 14.31]	91	<0.05
	Swelling or Hardness	2	10.00 [2.11, 47.24]	93	<0.05	3	7.71 [4.33, 13.72]	78	<0.05
	Fever	2	15.28 [10.11, 23.11	4	<0.05	3	5.07 [1.14, 22.44]	97	<0.05
	Headache	2	2.50 [2.14, 2.91]	79	<0.05	2	1.66 [1.35, 2.04]	77	<0.05
	Fatigue	2	2.47 [2.20, 2.78]	62	<0.05	2	1.75 [1.55, 1.98]	55	<0.05
	Myalgia	2	3.80 [3.35, 4.31]	0	<0.05	2	2.11 [1.41, 3.16]	81	<0.05
	Arthralgia	2	3.14 [2.68, 3.66]	0	<0.05	2	1.57 [1.11, 2.22]	57	<0.05
	Nausea or Vomiting	2	2.78 [2.31, 3.36]	0	<0.05	2	2.33 [1.92, 2.82]	0	<0.05
	Chills	2	5.85 [5.03, 6.79]	0	<0.05	2	4.37 [3.14, 6.09]	72	<0.05

Note: RR, Risk ratio; CI, confidence interval; $p < 0.05$.

Table 7. Specific adverse reactions in mRNA vaccine recipients aged 6–23 months versus two to five years.

		After Dose 1			After Dose 2		
	No. of Studies	RR (95% CI)	I^2	*p*-Value	RR (95% CI)	I^2	*p*-Value
Overall	1	0.74 [0.71, 0.77]	97	<0.05	0.80 [0.77, 0.83]	98	<0.05
Any local adverse reactions	1	0.70 [0.66, 0.74]	Not applicable	<0.05	0.73 [0.70, 0.77]	Not applicable	<0.05
Local pain	1	0.60 [0.57, 0.64]	Not applicable	<0.05	0.64 [0.60, 0.67]	Not applicable	<0.05
Erythema or Redness	1	1.53 [1.24, 1.89]	Not applicable	<0.05	1.53 [1.29, 1.80]	Not applicable	<0.05
Swelling or Hardness	1	1.85 [1.49, 2.31]	Not applicable	<0.05	1.83 [1.55, 2.15]	Not applicable	<0.05
Axillary swelling	1	0.91 [0.73, 1.13]	Not applicable	>0.05	1.02 [0.85, 1.23]	Not applicable	>0.05

Note: RR, Risk ratio; CI, confidence interval; $p < 0.05$.

3.2.3. Adverse Reactions to Different Dose Groups

In the mRNA and inactivated vaccine groups, participants received inconsistent doses. Also, participants in both the subunit vaccine and adenovirus vaccine groups received the same dose, so subgroup analysis failed to be performed on this basis.

3.3. *Immunogenicity of the COVID-19 Vaccines*

A total of 12 studies on the immunogenicity of COVID-19 vaccines were included in this systematic review article. Seven RCTs met the noninferiority of the immune response (detail in Table 8). In particular, Ali et al. also showed a GMR of 1.09 (95% CI: 0.94–1.26) for RBD-binding ELISA antibodies in adolescents aged 12–17 years relative to young adults aged 16–25. In addition, Thuluva et al. reported the GMT of 1099 in adolescents aged 12–17 years and 1148 in 5–12 years, with a neutralizing antibody GMR of 0.82 in 12–18 years and 0.86 in 5–12 years relative to adults, which meet the noninferiority criterion (i.e., the lower limit of two-sided 95% CI of GMT ratio is ≥0.5 limit set) with subunit vaccine as the intervention compared immune responses 14 days after booster vaccination in vaccinees and adults.

3.3.1. Humoral Immune Responses in Different Doses

Five RCTs provided data on seroconversion, which showed that the seroconversion after inoculation was significant, especially after the third dose (RR 392.95, 95%CI 24.66–6260.89, $p < 0.0001$; Figure S16 and Table 9) in inactivated vaccine groups. We found an increase in neutralizing antibodies as the number of doses increased in the inactivated vaccine and the adenovirus vaccine groups (Table 9). In addition, the data showed that the neutralizing antibody was significantly increased in 28 days after dose 2 in the subunit vaccine group. The result of Zhu et al. showed the seroconversion rate of RBD-binding antibodies 28 days after dose2 (RR 101.50, 95%CI 6.44–1600.76, $p = 0.001$, Figure S16 and Table 10) was higher than dose 1 (RR 99.48, 95%CI 6.31–1569.12, $p = 0.001$, Figure S16 and Table 10) in the adenovirus vector vaccine group, and the seroconversion rate of RBD-binding antibodies reached 100% in 28 days after booster vaccination.

Table 8. Immunogenicity of included studies.

Included Studies	Type of Vaccine	Days	Age (Years)	Dose	Participants	GMT (IU/mL)	GMR	Serologic Response	The Difference in Serologic Response	Noninferiority
Ali et al. [31]	mRNA-1273 vaccine (mRNA vaccine)	57	12–17	100 µg	340	1401.7 (1276.3, 1539.5)	1.08 (0.94, 1.25)	336/340 (97.0, 99.7)	0.2 (−1.8, 2.4)	Yes
			18–25	NA	296	1301.3 (1177.0, 1438.8)		292/296 (96.6, 99.6)		
Anderson et al. [32]	mRNA-1273 vaccine (mRNA vaccine)	57	6–23 months	25 µg	230	1781 (1616, 1962)	1.3 (1.1, 1.5)	230/230 (98.4, 100.0)	0.7 (−1.0, 2.5)	Yes
			2–5	25 µg	264	1410 (1272, 1563)	1.0 (0.9, 1.2)	261/264 (96.7, 99.8)	−0.4 (−2.7, 1.5)	Yes
			18–25	100 µg	294	1391 (1263, 1531)	/	289/291 (97.5, 99.9)	/	/
Áñez et al. [33]	NVX-CoV2373 (subunit vaccine)	35	12–17	0.5 mL	390	3860 (3423, 4352)	1.5 (1.3, 1.7)	-/390 (98.7%) (97.0, 99.6)	−1.0 (−2.8, 0.2)	Yes
			18–25	NA	416	2634 (2398, 2904)		-/416 (99.8%) (98.7, 100)		
Buddy Creech et al. [34]	mRNA-1273 vaccine (mRNA vaccine)	57	6–11	50 µg	320	1610.2 (1456.6, 1780.0)	1.2 (1.1, 1.4)	313/316 (97.3, 99.8)	0.1 (−1.9, 2.1)	Yes
			18–25	100 µg	295	1299.9 (1170.6, 1443.4)		292/295 (97.1, 99.8)		
Frenck et al. [35]	BNT162b2 Vaccine (mRNA vaccine)	A month after dose 2	12–15	30 µg	190	1239.5 (1095.5, 1402.5)	1.76 (1.47, 2.10)	NA	NA	Yes
			16–25	NA	170	705.1 (621.4, 800.2)		NA		
Liu et al. [37]	MVC-COV1901 (subunit vaccine)	57	12–17	0.5 mL	334	648.47 (608.62, 690.93)	1.16 (1.04, 1.29)	334/334 (98.90, 100.00)	−0.0% (0.00, 0.00)	Yes
			20–30	NA	210	559.54 (512.05, 611.34)		210/210 (98.26, 100.00)		
Walter et al. [39]	BNT162b2 vaccine (mRNA vaccine)	A month after dose 2	5–11	10 µg	264	1197.6 (1106.1, 1296.6)	1.04 (0.93, 1.18)	NA	NA	Yes
			16–25	30 µg	253	1146.5 (1045.5, 1257.2)		NA		

Note: GMT, geometric mean titers; GMR, geometric mean ratio; NA, not available.

Table 9. Neutralizing antibody in the vaccine groups versus the control groups.

Vaccine Type	Time	No. of Studies	RR (95% CI)	I^2	*p*-Value
Inactivated vaccine	28 days after dose 1	2	245.69 [34.93, 1727.83]	67	<0.05
	28 days after dose 2	3	363.09 [73.82, 1785.92]	0	<0.05
	28 days after dose 3	1	392.95 [24.66, 6260.89]	Not applicable	<0.05
Subunit vaccine	28 days after dose 2	1	31.29 [6.48, 151.07]	Not applicable	<0.05
Adenovirus vaccine	28 days after dose 1	1	14.67 [4.88, 44.04]	Not applicable	<0.05
	28 days after dose 2	1	24.50 [6.30, 95.28]	Not applicable	<0.05

Note: RR, Risk ratio; CI, confidence interval; $p < 0.05$.

Table 10. RBD-binding enzyme immunosorbent assay antibody in the adenovirus vaccine group.

	No. of Studies	RR (95% CI)	I^2	*p*-Value
28 days after Dose 1	1	99.48 [6.31, 1569.12]	Not applicable	<0.05
28 days after Dose 2	1	101.50 [6.44, 1600.76]	Not applicable	<0.05

Note: RR, Risk ratio; CI, confidence interval; $p < 0.05$.

3.3.2. Humoral Immune Responses in Different Ages

Publication bias was performed by funnel plot (Egger's test, $p = 0.026$). Subgroup analysis was performed because three RCTs provided seroconversions for different age groups at 28 days after vaccination. The data showed a significant humoral immune response to SARS-CoV-2 after receiving vaccination in all age groups, but the response appears to be inversely proportional to age, children aged three to five years (RR 125.90, 95%CI 25.72–616.35, $p < 0.00001$; Figure S16 and Table 11) have the most robust immune response of the three age groups at 28 days after the second dose. Similar differences were observed after the third dose; the response appears to be relatively high in children aged three to five years (RR 163.67, 95%CI 10.32–2594.58, $p = 0.0003$; Figure S16 and Table 11).

Table 11. Neutralizing antibody in the inactivated vaccine groups versus the control groups.

	No. of Studies	RR (95% CI)	I^2	p-Value
Neutralizing antibody 28 days after Dose 2				
3–5 years old	3	125.90 [25.72, 616.35]	0	<0.05
6–11/12 years old	3	122.82 [25.05, 602.16]	0	<0.05
12/13–17 years old	3	117.87 [24.04, 577.88]	0	<0.05
Neutralizing antibody 28 days after Dose 3				
3–5 years old	1	163.67 [10.32, 2594.58]	Not applicable	<0.05
6–12 years old	1	120.30 [7.61, 1901.57]	Not applicable	<0.05
13–17 years old	1	112.80 [7.13, 1783.53]	Not applicable	<0.05

Note: RR, Risk ratio; CI, confidence interval; $p < 0.05$.

3.3.3. Cellular Immune Responses

Two RCTs also assessed the ability of the COVID-19 vaccines to induce T-cell-mediated immunity among participants. Thuluva et al. showed that Th1 significantly skewed cellular immune response after CORBEVAX™ vaccination. Similarly, in the trial of Zhu et al., the data showed that a specific T-cell response was induced at day 28 after primary vaccination, particularly in Th1 cell responses.

3.4. *Efficacy of the COVID-19 Vaccines*

Among the studies on the efficacy of the COVID-19 vaccines, five RCTs were about the mRNA vaccine, three on the mRNA-1273 vaccine, and two were on the BNT162b2 vaccine, with about 100.0% (95% CI: 28.9%-NE%) efficacy was found in Ali et al., 36.8% (12.5% to 54.0%) of 2–5 years old and 50.6% (21.4% to 68.6%) of 6–23 months in Anderson et al., 88.0% (70.0–95.8%) in Buddy Creech et al., 100% (95% CI: 75.3–100%) in Frenck et al., 90.7% (95% CI: 67.4%–98.3%) in Walter et al., and one RCT was on subunit vaccine, about 79.5% (95% CI, 46.8% to 92.1%) efficacy was demonstrated against the predominant circulating Delta variant, in addition, 82.0% (95% CI, 32.4% to 95.2%) efficacy was found due to the SARS-CoV-2 delta variant. Both mRNA vaccines provided satisfactory prevention against COVID-19, especially seven days after the booster dose (RR 0.08, 95%CI 0.03–0.24, $p < 0.00001$; Figure S17 and Table 12). Other RCT studies with inactivated, subunit, or adenovirus vector vaccines as interventions did not evaluate the vaccine efficacy.

Table 12. COVID-19 was diagnosed after vaccination in the vaccine group versus the control group.

	No. of Studies	RR (95% CI)	I^2	p-Value
COVID-19 after the vaccination				
After dose 1 to before dose 2	2	0.16 [0.08, 0.32]	0	<0.05
Within 7 days after dose 2	1	0.09 [0.01, 1.64]	Not applicable	>0.05
7 days after dose 2	2	0.08 [0.03, 0.24]	0	<0.05
14 days after dose 2	3	0.30 [0.09, 0.97]	62	<0.05
COVID-19 after dose 2				
mRNA-1273 vaccine	3	0.30 [0.09, 0.97]	62	<0.05
BNT162b2 COVID-19 Vaccine	2	0.08 [0.03, 0.24]	0	<0.05

Note: RR, Risk ratio; CI, confidence interval; $p < 0.05$.

4. Discussion

As the global epidemic spreads, vaccinating children against COVID-19 has become one of the effective measures to prevent the development of the epidemic, but whether children are vaccinated largely depends on the wishes of parents or guardians. Parents with vaccine-hesitant were less knowledgeable about vaccines, the primary reason for concern is the vaccine safety and efficacy [41–43]. The children's age and current physical condition are other consideration factors for parents on vaccination, and parents are reluctant to vaccinate younger children and those who have been sick recently [44]. Other factors influence parents' vaccination intention and uptakes, such as parents' age, education, occupation, previous COVID-19 infection, and vaccination status [45,46]. Recent research shows that a high prevalence of severe COVID-19 was in children with comorbidities, such as obesity, diabetes, heart disease, and chronic lung diseases, and that neonate and premature infants also had a high risk [47]. Therefore, vaccination is vital.

An evaluation of COVID-19 vaccines will eliminate parents' doubts about vaccines and contribute to children's physical and mental health and all-around development. The findings of our review provide a comprehensive evidence profile on the safety, immunogenicity, and efficacy of COVID-19 vaccines in children younger than 18 years.

Our results show that the most common adverse reactions included local pain, swelling/hardness, and fever after the initial vaccination, and local pain, erythema/redness, swelling/hardness, and fever after the booster vaccination. Still, most local, and systemic adverse reactions were predominantly mild to moderate in severity and transient. Different from Du et al. [48], our meta-analysis found that the adenovirus vaccine was of the lowest safety, while the subunit vaccine was highest in our analyses of the four COVID-19 vaccines; this may be related to our inclusion of the subunit vaccine. However, the RCT [40] with adenoviral vector vaccine as an intervention was a small-sample study, and future studies are still required. In addition, there were no significant differences in total, systemic and local adverse reactions among different dose groups for various vaccines. Our results indicated that younger children may have a greater safety profile in the mRNA vaccine group and the inactivated vaccine group.

Good immunogenicity was observed in the included vaccine types. We found that the immune response to the mRNA and the subunit vaccine in adolescents was non-inferior in young people, consistent with the previous systemic review [49]. It was found that the humoral immune response is proportional to the number of doses in the inactivated and the adenovirus vaccine groups in our meta-analysis. In addition, our analysis found dose-level-dependent immunogenicity in the inactivated vaccine, which was in line with a newly published meta-analysis conducted by Du et al. [48]. The data of Han et al. showed that the higher dose of the vaccine could induce stronger immune responses in all age groups compared with the lower dose of the adenovirus vector vaccine. However, there are some different findings in our analysis; the immunogenicity's strength was negatively correlated with age in the inactivated vaccine. Some possibilities have been suggested. Other vaccines given to children produce a strong immune response to provide a better immune environment and generate cross-reactivity among the different beta coronaviruses, which may confer a nonspecific protective effect against SARS-CoV-2, such as measles, mumps, and rubella [50–52]. The immunity responses decreased with aging, indicating that a booster vaccine may be needed. However, the data showed a lower seroconversion rate and neutralizing antibody titer on day 28 in younger children (three to five years) than that of other age cohorts by Xia et al. [30]; future studies are still required to explore this result. The mRNA vaccines and subunit vaccines also elicited robust binding antibody responses to the prototype SARS-CoV-2, as well as against more recent variants: Alpha, Beta, Delta, and Omicron, including subvariants BA.1, BA.2, and BA.5 and the B.1.351 (beta), B.1.617.2 (delta), and Omicron variants. The data from Thuluva et al. showed the cellular immune response in the pediatric population demonstrated the expected Th1 skew. However, the specific T-cell response was not enhanced after booster vaccination by Zhu et al. [40].

Our analysis showed that the mRNA and subunit vaccines provided satisfactory protection against prototype SARS-CoV-2 and more recent variants. It should be especially noted that the research of Anderson et al. showed a lower efficacy, in which B.1.1.529 (Omicron) was the predominant circulating variant at the time of this experiment. The effectiveness of the mRNA vaccine declined during the Omicron period, and a similar phenomenon was also observed in other research on children and adults [53–55]. Like the previous studies, the vaccines still have a protective effect even during an epidemic of a new variant [56–58], which also led to significant heterogeneity in effectiveness analyses, and there was little change in the meta-analysis result with Anderson's RCT removed, indicating that the analysis results were robust.

Compared with the previous meta-analysis, this is the first meta-analysis to include children aged six months to three years old on COVID-19 vaccines. In addition, a new type of vaccine (subunit vaccine) has been added to our analysis, providing a more comprehensive assessment of existing vaccines' safety, immunogenicity, and efficacy. Additionally, our review evaluates the effectiveness of the mRNA vaccine against the Omicron variants. This review included the latest high-quality randomized controlled studies and had a large sample size, with 17,731 participants in the experimental group and 7444 in the control group, which provides strong evidence for vaccine evaluation.

There are several limitations in our systematic review and meta-analysis. First, there is a lack of data on younger children under six months and a lack of longer-term follow-up to assess the duration of immune responses, efficacy, and safety for children younger than 18. In addition, our analysis included four types of COVID-19 vaccines (the mRNA, inactivated, subunit, and adenovirus vector vaccine); however, there was just one RCT about the adenoviral vector vaccine as an intervention with a small sample, and only two RCTs provided relevant data on cellular immune responses in our analysis. Further studies are still required. In addition, our meta-analysis did not evaluate the COVID-19 vaccines in high-risk children, nor did we evaluate the effectiveness of the COVID-19 vaccine by hospitalization, severe illness, and mortality rates of children in the vaccine and control groups, due to limited data. Last and most importantly, high unexplained heterogeneity could be found in some subgroups in our review, which might be attributed to the variation in different variants, the design of studies, vaccine dose, sociodemographic factors, etc. Therefore, the safety, immunogenicity, and efficacy of different COVID-19 vaccines in children younger than 18 years, especially under six months, still require extensive and high-quality studies and longer follow-up periods.

The following questions remain about the vaccination of children under the age of 18. First, long-term follow-up is needed to assess the duration and efficacy of the immune response to COVID-19 vaccines. Second, it is urgent to evaluate whether the COVID-19 vaccines cause severe side effects such as glomerulonephritis, myocarditis, and chronic fatigue syndrome. Finally, more attention should be given to vaccinating high-risk children to protect them from contracting COVID-19.

5. Conclusions

Based on the systematic analysis of the four COVID-19 vaccines, we found that the four vaccines are generally safe and feasible with no serious side effects, but considering that some vaccines have been less studied, further research is needed. The immunogenicity and effectiveness of the four vaccines in children younger than 18 years are acceptable and approved, which may improve parents' confidence in COVID-19 vaccinations. However, there are no data on children younger than six months, and more research are needed. Longer-term follow-up is required to assess the duration of immune responses, efficacy, and safety.

Supplementary Materials: The following supporting information can be downloaded at: https://www.mdpi.com/article/10.3390/vaccines11010087/s1, Figure S1. Funnel plot for the total adverse reactions; Figure S2. Funnel plot for the systemic and local adverse reactions; Figure S3. Funnel plot for the neutralizing antibody 28 days after dose 2; Figure S4. Adverse reactions among vaccination group versus control group; Figure S5. Specific adverse reactions in the mRNA vaccine group versus the control group; Figure S6. Specific adverse reactions in the inactivated vaccine group versus the control group; Figure S7. Specific adverse reactions in the subunit vaccine group versus the control group; Figure S8. Specific adverse reactions in the adenovirus vector vaccine group versus the control group; Figure S9. Adverse reactions in the mRNA vaccine group of different ages versus the control group; Figure S10. Specific adverse reactions in the mRNA vaccine recipients aged 12–15 years versus 5–11 years; Figure S11. Specific adverse reactions in the mRNA vaccine recipients aged 6–23 months versus 2–5 years; Figure S12. Adverse reactions in the inactivated vaccine group of different ages versus the control group after whole vaccination; Figure S13. Specific adverse reactions in the inactivated vaccine recipients aged 12–17 years versus 3–12 years; Figure S14. Adverse reactions in the subunit vaccine group within 7 days of different ages versus the control group after whole vaccination; Figure S15. Specific adverse reactions in the subunit vaccine recipients aged 12–17 years versus 5–11 years; Figure S16. Seroconversion rate in the vaccine group versus the control group; Figure S17. COVID-19 diagnosed after vaccination in the vaccine group versus the control group; Table S1. Search formula in PubMed; Table S2. Search formula in Embase; Table S3. Search formula in the Cochrane library; Table S4. Specific adverse reactions in the inactivated vaccine group versus the control group after dose 1 and dose 2; Table S5. Specific adverse reactions in the subunit vaccine group versus the control group after dose 1 and dose 2; Table S6. Specific adverse reactions in the adenovirus vector vaccine group versus the control group after dose 1 and dose 2; Table S7. Specific adverse reactions in inactivated vaccine recipients of different ages after whole vaccinations; Table S8. Specific adverse reactions in subunit vaccine recipients of different ages after whole vaccinations.

Author Contributions: Conceived and designed the study: Y.S. and Y.T.; Retrieval, selection, and extraction: Y.T. and L.C.; Analyzed the data: Y.T. and L.C.; Drafting of the manuscript: Y.T.; Critical revision: Y.S. and Y.T.; Supervision: Y.S. All authors have read and agreed to the published version of the manuscript.

Funding: National Key Research and Development Program of China (2022YFC2704803).

Institutional Review Board Statement: The manuscript is a systematic review; no ethics approval was required.

Informed Consent Statement: Not applicable.

Data Availability Statement: All data are included and/or available within the Supplementary Materials. Additional data about this paper may be requested from the authors.

Conflicts of Interest: The authors declare no conflict of interest.

References

1. Lee, P.-I.; Hu, Y.-L.; Chen, P.-Y.; Huang, Y.-C.; Hsueh, P.-R. Are children less susceptible to COVID-19? *J. Microbiol. Immunol. Infect.* **2020**, *53*, 371–372. [CrossRef] [PubMed]
2. Wang, C.; Tee, M.; Roy, A.E.; Fardin, M.A.; Srichokchatchawan, W.; Habib, H.A.; Tran, B.X.; Hussain, S.; Hoang, M.T.; Le, X.T.; et al. The impact of COVID-19 pandemic on physical and mental health of Asians: A study of seven middle-income countries in Asia. *PLoS ONE* **2021**, *16*, e0246824. [CrossRef] [PubMed]
3. Wang, Z.; Zhou, Q.; Wang, C.; Shi, Q.; Lu, S.; Ma, Y.; Luo, X.; Xun, Y.; Li, W.; Baskota, M.; et al. Clinical characteristics of children with COVID-19: A rapid review and meta-analysis. *Ann. Transl. Med.* **2020**, *8*, 620. [CrossRef]
4. de Souza, T.H.; Nadal, J.A.; Nogueira, R.J.N.; Pereira, R.M.; Brandão, M.B. Clinical manifestations of children with COVID-19: A systematic review. *Pediatr. Pulmonol.* **2020**, *55*, 1892–1899. [CrossRef] [PubMed]
5. Zimmermann, P.; Curtis, N. Why is COVID-19 less severe in children? A review of the proposed mechanisms underlying the age-related difference in severity of SARS-CoV-2 infections. *Arch. Dis. Child.* **2021**, *106*, 429–439. [CrossRef]
6. Havers, F.P.; Whitaker, M.; Self, J.L.; Chai, S.J.; Kirley, P.D.; Alden, N.B.; Kawasaki, B.; Meek, J.; Yousey-Hindes, K.; Anderson, E.J.; et al. Hospitalization of Adolescents Aged 12–17 Years with Laboratory-Confirmed COVID-19—COVID-NET, 14 States, March 1, 2020–April 24, 2021. *MMWR. Morb. Mortal. Wkly. Rep.* **2021**, *70*, 851–857. [CrossRef]
7. Jiang, L.; Tang, K.; Levin, M.; Irfan, O.; Morris, S.K.; Wilson, K.; Klein, J.D.; Bhutta, Z.A. COVID-19 and multisystem inflammatory syndrome in children and adolescents. *Lancet Infect. Dis.* **2020**, *20*, e276–e288. [CrossRef]

8. Penner, J.; Abdel-Mannan, O.; Grant, K.; Maillard, S.; Kucera, F.; Hassell, J.; Eyre, M.; Berger, Z.; Hacohen, Y.; Moshal, K.; et al. 6-month multidisciplinary follow-up and outcomes of patients with paediatric inflammatory multisystem syndrome (PIMS-TS) at a UK tertiary paediatric hospital: A retrospective cohort study. *Lancet Child Adolesc. Health* **2021**, *5*, 473–482. [CrossRef]
9. Townsend, E. Debate: The impact of school closures and lockdown on mental health in young people. *Child Adolesc. Ment. Health* **2020**, *25*, 265–266. [CrossRef]
10. O'Sullivan, K.; Clark, S.; McGrane, A.; Rock, N.; Burke, L.; Boyle, N.; Joksimovic, N.; Marshall, K. A Qualitative Study of Child and Adolescent Mental Health during the COVID-19 Pandemic in Ireland. *Int. J. Environ. Res. Public Health* **2021**, *18*, 1062. [CrossRef]
11. Tang, S.; Xiang, M.; Cheung, T.; Xiang, Y.-T. Mental health and its correlates among children and adolescents during COVID-19 school closure: The importance of parent-child discussion. *J. Affect. Disord.* **2020**, *279*, 353–360. [CrossRef] [PubMed]
12. Ma, L.; Mazidi, M.; Li, K.; Li, Y.; Chen, S.; Kirwan, R.; Zhou, H.; Yan, N.; Rahman, A.; Wang, W.; et al. Prevalence of mental health problems among children and adolescents during the COVID-19 pandemic: A systematic review and meta-analysis. *J. Affect. Disord.* **2021**, *293*, 78–89. [CrossRef] [PubMed]
13. de Oliveira, J.M.D.; Butini, L.; Pauletto, P.; Lehmkuhl, K.M.; Stefani, C.M.; Bolan, M.; Guerra, E.; Dick, B.; Canto, G.D.L.; Massignan, C. Mental health effects prevalence in children and adolescents during the COVID-19 pandemic: A systematic review. *Worldviews Evid.-Based Nurs.* **2022**, *19*, 130–137. [CrossRef]
14. McKune, S.L.; Acosta, D.; Diaz, N.; Brittain, K.; Beaulieu, D.J.; Maurelli, A.T.; Nelson, E.J. Psychosocial health of school-aged children during the initial COVID-19 safer-at-home school mandates in Florida: A cross-sectional study. *BMC Public Health* **2021**, *21*, 603. [CrossRef] [PubMed]
15. Engzell, P.; Frey, A.; Verhagen, M.D. Learning loss due to school closures during the COVID-19 pandemic. *Proc. Natl. Acad. Sci. USA* **2021**, *118*, e2022376118. [CrossRef]
16. Larsen, L.; Helland, M.S.; Holt, T. The impact of school closure and social isolation on children in vulnerable families during COVID-19: A focus on children's reactions. *Eur. Child Adolesc. Psychiatry* **2021**, *31*, 1–11. [CrossRef]
17. Rajmil, L.; Hjern, A.; Boran, P.; Gunnlaugsson, G.; de Camargo, O.K.; Raman, S. Impact of lockdown and school closure on children's health and well-being during the first wave of COVID-19: A narrative review. *BMJ Paediatr. Open* **2021**, *5*, e001043. [CrossRef]
18. Parks, S.E.; Zviedrite, N.; Budzyn, S.E.; Panaggio, M.J.; Raible, E.; Papazian, M.; Magid, J.; Ahmed, F.; Uzicanin, A.; Barrios, L.C. COVID-19–Related School Closures and Learning Modality Changes—United States, August 1–September 17, 2021. *MMWR. Morb. Mortal. Wkly. Rep.* **2021**, *70*, 1374–1376. [CrossRef]
19. Dembiński, Ł.; Huss, G.; Radziewicz-Winnicki, I.; Grossman, Z.; Mazur, A.; del Torso, S.; Barak, S.; Sanz, A.C.; Hadjipanayis, A. EAP and ECPCP Statement Risks for Children's Health During the COVID-19 Pandemic and a Call for Maintenance of Essential Pediatric Services. *Front. Pediatr.* **2021**, *9*, 679803. [CrossRef]
20. Chun, J.Y.; Jeong, H.; Kim, Y. Identifying susceptibility of children and adolescents to the Omicron variant (B.1.1.529). *BMC Med.* **2022**, *20*, 451. [CrossRef]
21. Goldman, E. How the unvaccinated threaten the vaccinated for COVID-19: A Darwinian perspective. *Proc. Natl. Acad. Sci. USA* **2021**, *118*, e2114279118. [CrossRef] [PubMed]
22. Tian, F.; Yang, R.; Chen, Z. Safety and efficacy of COVID-19 vaccines in children and adolescents: A systematic review of randomized controlled trials. *J. Med. Virol.* **2022**, *94*, 4644–4653. [CrossRef] [PubMed]
23. Cauchemez, S.; Bosetti, P.; Kiem, C.T.; Mouro, V.; Consoli, A.; Fontanet, A. Education and mental health: Good reasons to vaccinate children. *Lancet* **2021**, *398*, 387. [CrossRef] [PubMed]
24. Klass, P.; Ratner, A.J. Vaccinating Children against Covid-19—The Lessons of Measles. *New Engl. J. Med.* **2021**, *384*, 589–591. [CrossRef] [PubMed]
25. Anderson, E.J.; Campbell, J.D.; Creech, C.B.; Frenck, R.; Kamidani, S.; Munoz, F.M.; Nachman, S.; Spearman, P. Warp Speed for Coronavirus Disease 2019 (COVID-19) Vaccines: Why Are Children Stuck in Neutral? *Clin. Infect. Dis.* **2020**, *73*, 336–340. [CrossRef]
26. Opel, D.J.; Diekema, D.S.; Ross, L.F. Should We Mandate a COVID-19 Vaccine for Children? *JAMA Pediatr.* **2021**, *175*, 125–126. [CrossRef]
27. Thunström, L.; Ashworth, M.; Finnoff, D.; Newbold, S.C. Hesitancy Toward a COVID-19 Vaccine. *EcoHealth* **2021**, *18*, 44–60. [CrossRef]
28. Babicki, M.; Pokorna-Kałwak, D.; Doniec, Z.; Mastalerz-Migas, A. Attitudes of Parents with Regard to Vaccination of Children against COVID-19 in Poland. A Nationwide Online Survey. *Vaccines* **2021**, *9*, 1192. [CrossRef]
29. Xia, S.; Duan, K.; Zhang, Y.; Zeng, X.; Zhao, D.; Zhang, H.; Xie, Z.; Li, X.; Peng, C.; Zhang, W.; et al. Safety and Immunogenicity of an Inactivated COVID-19 Vaccine, WIBP-CorV, in Healthy Children: Interim Analysis of a Randomized, Double-Blind, Controlled, Phase 1/2 Trial. *Front. Immunol.* **2022**, *13*, 196–208. [CrossRef]
30. Xia, S.; Zhang, Y.; Wang, Y.; Wang, H.; Yang, Y.; Gao, G.F.; Tan, W.; Wu, G.; Xu, M.; Lou, Z.; et al. Safety and immunogenicity of an inactivated COVID-19 vaccine, BBIBP-CorV, in people younger than 18 years: A randomised, double-blind, controlled, phase 1/2 trial. *Lancet Infect. Dis.* **2022**, *22*, 196–208. [CrossRef]
31. Ali, K.; Berman, G.; Zhou, H.; Deng, W.; Faughnan, V.; Coronado-Voges, M.; Ding, B.; Dooley, J.; Girard, B.; Hillebrand, W.; et al. Evaluation of mRNA-1273 SARS-CoV-2 Vaccine in Adolescents. *N. Engl. J. Med.* **2021**, *385*, 2241–2251. [CrossRef]

32. Anderson, E.J.; Creech, C.B.; Berthaud, V.; Piramzadian, A.; Johnson, K.A.; Zervos, M.; Garner, F.; Griffin, C.; Palanpurwala, K.; Turner, M.; et al. Evaluation of mRNA-1273 Vaccine in Children 6 Months to 5 Years of Age. *N. Engl. J. Med.* **2022**, *387*, 1673–1687. [CrossRef] [PubMed]
33. Áñez, G.; Dunkle, L.M.; Gay, C.L.; Kotloff, K.L.; Adelglass, J.M.; Essink, B.; Campbell, J.D.; Cloney-Clark, S.; Zhu, M.; Plested, J.S.; et al. Safety, Immunogenicity and Efficacy of NVX-CoV2373 in Adolescents in PREVENT-19: A Randomized, Phase 3 Trial. *medRxiv* **2022**. [CrossRef]
34. Creech, C.B.; Anderson, E.; Berthaud, V.; Yildirim, I.; Atz, A.M.; Baez, I.M.; Finkelstein, D.; Pickrell, P.; Kirstein, J.; Yut, C.; et al. Evaluation of mRNA-1273 Covid-19 Vaccine in Children 6 to 11 Years of Age. *N. Engl. J. Med.* **2022**, *386*, 2011–2023. [CrossRef] [PubMed]
35. Frenck, R.W., Jr.; Klein, N.P.; Kitchin, N.; Gurtman, A.; Absalon, J.; Lockhart, S.; Perez, J.L.; Walter, E.B.; Senders, S.; Bailey, R.; et al. Safety, Immunogenicity, and Efficacy of the BNT162b2 Covid-19 Vaccine in Adolescents. *N. Engl. J. Med.* **2021**, *385*, 239–250. [CrossRef] [PubMed]
36. Han, B.; Song, Y.; Li, C.; Yang, W.; Ma, Q.; Jiang, Z.; Li, M.; Lian, X.; Jiao, W.; Wang, L.; et al. Safety, tolerability, and immunogenicity of an inactivated SARS-CoV-2 vaccine (CoronaVac) in healthy children and adolescents: A double-blind, randomised, controlled, phase 1/2 clinical trial. *Lancet Infect. Dis.* **2021**, *21*, 1645–1653. [CrossRef] [PubMed]
37. Liu, L.T.C.; Chiu, C.H.; Chiu, N.C.; Tan, B.F.; Lin, C.Y.; Cheng, H.Y.; Lin, M.Y.; Lien, C.E.; Chen, C.; Huang, L.M. Safety and immunogenicity of SARS-CoV-2 vaccine MVC-COV1901 in adolescents in Taiwan: A double-blind, randomized, placebo-controlled phase 2 trial. *medRxiv* **2022**. [CrossRef]
38. Thuluva, S.; Paradkar, V.; Gunneri, S.; Yerroju, V.; Mogulla, R.R.; Suneetha, P.V.; Turaga, K.; Kyasani, M.; Manoharan, S.K.; Adabala, S.; et al. Safety, tolerability and immunogenicity of Biological E's CORBEVAX™ vaccine in children and adolescents: A prospective, randomised, double-blind, placebo controlled, phase-2/3 study. *Vaccine* **2022**, *40*, 7130–7140. [CrossRef]
39. Walter, E.B.; Talaat, K.R.; Sabharwal, C.; Gurtman, A.; Lockhart, S.; Paulsen, G.C.; Barnett, E.D.; Muñoz, F.M.; Maldonado, Y.; Pahud, B.A.; et al. Evaluation of the BNT162b2 Covid-19 Vaccine in Children 5 to 11 Years of Age. *N. Engl. J. Med.* **2022**, *386*, 35–46. [CrossRef]
40. Zhu, F.; Jin, P.; Zhu, T.; Wang, W.; Ye, H.; Pan, H.; Hou, L.; Li, J.; Wang, X.; Wu, S.; et al. Safety and Immunogenicity of a Recombinant Adenovirus Type-5–Vectored Coronavirus Disease 2019 (COVID-19) Vaccine with a Homologous Prime-Boost Regimen in Healthy Participants Aged $\geq$6 Years: A Randomized, Double-Blind, Placebo-Controlled, Phase 2b Trial. *Clin. Infect. Dis.* **2021**, *75*, e783–e791. [CrossRef]
41. Ruiz, J.B.; Bell, R.A. Parental COVID-19 Vaccine Hesitancy in the United States. *Public Health Rep.* **2022**, *137*, 1162–1169. [CrossRef]
42. Russo, L.; Croci, I.; Campagna, I.; Pandolfi, E.; Villani, A.; Reale, A.; Barbieri, M.A.; Raponi, M.; Gesualdo, F.; Tozzi, A.E. Intention of Parents to Immunize Children against SARS-CoV-2 in Italy. *Vaccines* **2021**, *9*, 1469. [CrossRef]
43. Wang, Q.; Xiu, S.; Zhao, S.; Wang, J.; Han, Y.; Dong, S.; Huang, J.; Cui, T.; Yang, L.; Shi, N.; et al. Vaccine Hesitancy: COVID-19 and Influenza Vaccine Willingness among Parents in Wuxi, China—A Cross-Sectional Study. *Vaccines* **2021**, *9*, 342. [CrossRef] [PubMed]
44. Lu, L.; Gu, W.; Xie, H.; Wang, X.; Cao, L.; Shan, M.; Wu, P.; Tian, Y.; Zhou, K. Parental Attitudes Towards Vaccination Against COVID-19 in China During Pandemic. *Infect. Drug Resist.* **2022**, *15*, 4541–4546. [CrossRef]
45. Khatatbeh, M.; Albalas, S.; Khatatbeh, H.; Momani, W.; Melhem, O.; Al Omari, O.; Tarhini, Z.; A'Aqoulah, A.; Al-Jubouri, M.; Nashwan, A.J.; et al. Children's rates of COVID-19 vaccination as reported by parents, vaccine hesitancy, and determinants of COVID-19 vaccine uptake among children: A multi-country study from the Eastern Mediterranean Region. *BMC Public Health* **2022**, *22*, 1375. [CrossRef]
46. del Giudice, G.M.; Napoli, A.; Corea, F.; Folcarelli, L.; Angelillo, I.F. Evaluating COVID-19 Vaccine Willingness and Hesitancy among Parents of Children Aged 5–11 Years with Chronic Conditions in Italy. *Vaccines* **2022**, *10*, 396. [CrossRef] [PubMed]
47. Choi, J.H.; Choi, S.-H.; Yun, K.W. Risk Factors for Severe COVID-19 in Children: A Systematic Review and Meta-Analysis. *J. Korean Med. Sci.* **2022**, *37*, e35. [CrossRef]
48. Du, Y.; Chen, L.; Shi, Y. Safety, Immunogenicity, and Efficacy of COVID-19 Vaccines in Adolescents, Children, and Infants: A Systematic Review and Meta-Analysis. *Front. Public Health* **2022**, *10*, 829176. [CrossRef] [PubMed]
49. Lv, M.; Luo, X.; Shen, Q.; Lei, R.; Liu, X.; Liu, E.; Li, Q.; Chen, Y. Safety, Immunogenicity, and Efficacy of COVID-19 Vaccines in Children and Adolescents: A Systematic Review. *Vaccines* **2021**, *9*, 1102. [CrossRef]
50. Fidel, P.L.; Noverr, M.C. Could an Unrelated Live Attenuated Vaccine Serve as a Preventive Measure To Dampen Septic Inflammation Associated with COVID-19 Infection? *mBio* **2020**, *11*, e907–e920. [CrossRef] [PubMed]
51. Dhochak, N.; Singhal, T.; Kabra, S.K.; Lodha, R. Pathophysiology of COVID-19: Why Children Fare Better than Adults? *Indian J. Pediatr.* **2020**, *87*, 537–546. [CrossRef] [PubMed]
52. Piccaluga, P.P.; Malerba, G.; Navari, M.; Diani, E.; Concia, E.; Gibellini, D. Cross-Immunization Against Respiratory Coronaviruses May Protect Children From SARS-CoV2: More Than a Simple Hypothesis? *Front. Pediatr.* **2021**, *8*, 595539. [CrossRef] [PubMed]
53. Price, A.M.; Olson, S.M.; Newhams, M.M.; Halasa, N.B.; Boom, J.A.; Sahni, L.C.; Pannaraj, P.S.; Irby, K.; Bline, K.E.; Maddux, A.B.; et al. BNT162b2 Protection against the Omicron Variant in Children and Adolescents. *N. Engl. J. Med.* **2022**, *386*, 1899–1909. [CrossRef] [PubMed]
54. Tseng, H.F.; Ackerson, B.K.; Luo, Y.; Sy, L.S.; Talarico, C.A.; Tian, Y.; Bruxvoort, K.J.; Tubert, J.E.; Florea, A.; Ku, J.H.; et al. Effectiveness of mRNA-1273 against SARS-CoV-2 Omicron and Delta variants. *Nat. Med.* **2022**, *28*, 1063–1071. [CrossRef]

55. Andrews, N.; Stowe, J.; Kirsebom, F.; Toffa, S.; Rickeard, T.; Gallagher, E.; Gower, C.; Kall, M.; Groves, N.; O'Connell, A.-M.; et al. Covid-19 Vaccine Effectiveness against the Omicron (B.1.1.529) Variant. *N. Engl. J. Med.* **2022**, *386*, 1532–1546. [CrossRef] [PubMed]
56. Harris, R.J.; Hall, J.A.; Zaidi, A.; Andrews, N.J.; Dunbar, J.K.; Dabrera, G. Effect of Vaccination on Household Transmission of SARS-CoV-2 in England. *N. Engl. J. Med.* **2021**, *385*, 759–760. [CrossRef]
57. Prunas, O.; Warren, J.L.; Crawford, F.W.; Gazit, S.; Patalon, T.; Weinberger, D.M.; Pitzer, V.E. Vaccination with BNT162b2 reduces transmission of SARS-CoV-2 to household contacts in Israel. *Science* **2022**, *375*, 1151–1154. [CrossRef]
58. Fowlkes, A.L.; Yoon, S.K.; Lutrick, K.; Gwynn, L.; Burns, J.; Grant, L.; Phillips, A.L.; Ellingson, K.; Ferraris, M.V.; LeClair, L.B.; et al. Effectiveness of 2-Dose BNT162b2 (Pfizer BioNTech) mRNA Vaccine in Preventing SARS-CoV-2 Infection Among Children Aged 5–11 Years and Adolescents Aged 12–15 Years—PROTECT Cohort, July 2021–February 2022. *MMWR. Morb. Mortal. Wkly. Rep.* **2022**, *71*, 422–428. [CrossRef]

vaccines MDPI

Review

Review of Vaccination Recommendations in Guidelines for Non-Communicable Diseases with Highest Global Disease Burden among Adults 75 Years Old and Above

Abdul Rahman Ishak, Yu Chun Hsieh *, Harshitha Srinivasan and Kay Choong See *

Department of Medicine, Yong Loo Lin School of Medicine, National University of Singapore, Singapore 117597, Singapore; e0501186@u.nus.edu (A.R.I.); e0408893@u.nus.edu (H.S.)
* Correspondence: hsieh.yu.chun@u.nus.edu (Y.C.H.); kaychoongsee@nus.edu.sg (K.C.S.)

Abstract: This scientific review paper explores international and country-specific healthcare guidelines for non-communicable diseases with the highest burden among individuals aged 75 years and above. The study aims to identify the best vaccination practices and standardize healthcare practices to improve vaccination adherence in this vulnerable population. Given that older people are more prone to infectious illnesses and have higher rates of morbidity and mortality, vaccinations are essential for disease prevention. Despite the proven efficacy of vaccinations, adherence has plateaued in recent years, partly due to a lack of accessibility, public education, and variability in disease-specific guidelines. This paper highlights the need for a more robust and standardized international vaccination model to improve quality of life and reduce disability-adjusted life years among the elderly. The findings of this study call for further research to review the guidelines as more implementations are put in place, including non-English guidelines.

Keywords: vaccination; guidelines; geriatric; elderly; public health; review

check for updates

Citation: Ishak, A.R.; Hsieh, Y.C.; Srinivasan, H.; See, K.C. Review of Vaccination Recommendations in Guidelines for Non-Communicable Diseases with Highest Global Disease Burden among Adults 75 Years Old and Above. *Vaccines* **2023**, *11*, 1076. https://doi.org/10.3390/vaccines11061076

Academic Editor: Kevin Coombs

Received: 13 April 2023
Revised: 31 May 2023
Accepted: 6 June 2023
Published: 8 June 2023

1. Introduction

Development of vaccinations has been touted as one of the most cost-effective and efficacious public health interventions in the world. An estimated 2.5 million deaths are prevented by vaccinations each year [1]. Increased accessibility to vaccines is prolonging life expectancy and decreasing morbidity and mortality caused by vaccine-preventable diseases. Primary prevention through vaccinations has an integral part to play in caring for patients, especially in vulnerable populations such as the elderly. There is a notable direct relationship between increasing age and susceptibility to infections among the elderly, with infectious diseases accounting for one third of deaths in adults aged 65 and older [2,3].

Our project focuses on non-communicable diseases (NCD) which are defined by the Pan American Health Organization (PAHO) as a "group of conditions that are not mainly caused by an acute infection, result in long-term health consequences and often create a need for long-term treatment and care" [4]. According to the World Health Organization (WHO), NCDs are the leading cause of death worldwide, responsible for 71% of the total number of deaths each year [5].

Vaccination adherence has plateaued in recent years, largely due to the COVID-19 pandemic and its associated disruptions [6]. Other factors such as a lack of accessibility and public education also contribute to vaccination uptake being suboptimal.

One way to shape policy and allow for better vaccine implementation is publishing guideline recommendations. The WHO has a database with papers recommending vaccinations to prevent various life-threatening conditions [7].

Disease-specific guidelines are more relevant and influential in clinical practice. However, their vaccination recommendations may be variable, contributing to suboptimal vaccination practices. Therefore, there is a need to review the healthcare guidelines of different countries, identify the role of vaccinations in the proposed management, and standardise these healthcare practices in order to boost vaccination adherence.

This project aims to examine international and country-specific healthcare guidelines on non-communicable diseases with greatest burden as quantified by mortality, morbidity, and disability-adjusted life years among the age group of 75 years old and above. With the increasing vulnerability of the older age group, this study is pertinent to identify the best vaccination practices for a more robust and standardised international vaccination model to be established.

2. Materials and Methods

Please see Figure 1 for a summary of the Materials and Methods.

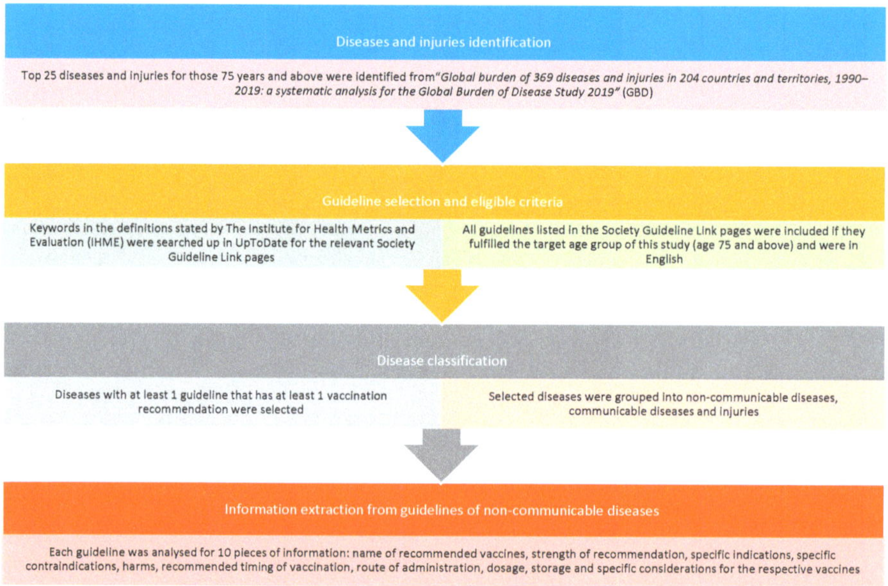

Figure 1. Flowchart of materials and methods.

2.1. Diseases and Injuries Identification

The study "Global burden of 369 diseases and injuries in 204 countries and territories, 1990–2019: a systematic analysis for the Global Burden of Disease Study 2019" (GBD) [8], was referenced for this study. The top 25 diseases and injuries for those 75 years and above, ranked according to disability-adjusted life years (DALYs), were identified.

The 25 identified diseases and injuries are as follows: ischaemic heart disease (IHD); stroke; chronic obstructive pulmonary disease (COPD); Alzheimer's disease and other dementias; diabetes; lower respiratory infections (LRI); tracheal, bronchus, and lung cancer; falls; chronic kidney disease (CKD); age-related hearing loss; hypertensive heart disease; diarrhoeal diseases; low back pain; colon and rectal cancer; blindness and vision loss; atrial fibrillation and flutter; stomach cancer; prostate cancer, cirrhosis and other chronic liver diseases; Parkinson's disease; osteoarthritis; oral disorders; tuberculosis (TB); asthma; and road injuries.

2.2. Guideline Selection and Eligible Criteria

The Institute for Health Metrics and Evaluation (IHME), which published GBD, has a definition for every disease and injury. The keywords mentioned in the definition were then searched in UpToDate for the relevant Society Guideline Link pages.

UpToDate (UpToDate Inc, Waltham, MA, USA) is an evidence-based point of care medical resource that is widely used by healthcare professionals [9] due to its feasibility and usefulness in clinical decision-making [10]; it was referenced for the list of guidelines.

Guidelines listed in the Society Guideline Link pages were included if they fulfilled the target age group of this study (age 75 and above) and were in English. The guideline selection process was undertaken from September 2022 to November 2022.

2.3. Disease Classification

Diseases with at least one guideline that has at least one vaccination recommendation were selected. This group of diseases was then classified into non-communicable diseases, communicable diseases and injuries. This classification system is used by WHO to analyse the composition of DALYs in the World Health Statistics 2023 [11].

2.4. Information Extraction

Each guideline of each non-communicable disease was further analysed for ten pieces of information: name of recommended vaccines, strength of recommendation, specific indications, specific contraindications, harms, recommended timing of vaccination, route of administration, dosage, storage, and specific considerations for the respective vaccines.

Strength of recommendation refers to the extent of the confidence the panel of experts have in a specific recommendation, usually following the analysis of the benefits, risks, the context, and the quality of the evidence [12]. The strength of recommendation influences the ease and/or complexity of adopting and implementing a recommendation [13].

Indications, contraindications, harms are crucial for healthcare professionals to weigh the risks and benefits of vaccination, as these affect their confidence in offering the vaccination to their patients [14,15].

3. Results

Out of the 25 identified diseases, 23 (ischaemic heart disease; stroke; chronic obstructive pulmonary disease; Alzheimer's disease and other dementias; diabetes; lower respiratory infections; tracheal, bronchus, and lung cancer; falls; chronic kidney disease; age-related hearing loss; hypertensive heart disease; diarrhoeal diseases; low back pain; colon and rectal cancer; atrial fibrillation and flutter; stomach cancer; prostate cancer; cirrhosis and other chronic liver diseases; Parkinson's disease; osteoarthritis; tuberculosis; asthma; and road injuries) contained guidelines listed in the Society Guideline Link on UpToDate. Upon further analysis of these 23 diseases, 15 (ischaemic heart disease; stroke; chronic obstructive pulmonary disease; diabetes; lower respiratory infections; falls; chronic kidney disease; hypertensive heart disease, diarrhoeal diseases; colon and rectal cancer; atrial fibrillation and flutter; stomach cancer; cirrhosis and other chronic liver diseases; tuberculosis; asthma) contained guidelines which mentioned vaccine recommendations.

Out of these 15 diseases, we will focus on 11 of the non-communicable diseases namely: ischaemic heart disease; stroke; chronic obstructive pulmonary disease; diabetes; chronic kidney disease; hypertensive heart disease; colon and rectal cancer; atrial fibrillation and flutter; stomach cancer; cirrhosis and other chronic liver diseases; and asthma). Lower respiratory infections, diarrhoeal diseases, and tuberculosis are excluded as they are communicable diseases. Falls are excluded because according to the World Health Organisation (WHO), they are grouped separately.

3.1. IHD

For ischaemic heart disease, 6 out of 102 guidelines mentioned vaccine recommendations. These guidelines were from the United States, Europe and Australia–New Zealand. All six guidelines recommended the influenza vaccine. Only one guideline, which was from Australia–New Zealand, recommended the pneumococcal vaccine.

This paragraph focuses on guidelines recommending the influenza vaccine. In terms of their strength of recommendation, three guidelines were strong, while three were not stated.

Additionally, specific indications from the six guidelines include stable ischaemic heart disease patients aged 65 years old and above; all CABG patients unless contraindications exist; chronic coronary syndrome patients aged 65 and above; and everyone with coronary heart disease unless contraindicated. Only one guideline mentioned specific harms, which were pain and myalgia at the injection site. Overall, it is recommended that the influenza vaccine be given annually via intramuscular route, using a standard dose (volume was not mentioned). There was also no mention of any storage information.

This paragraph focuses on the one guideline recommending the pneumococcal vaccine. The strength of recommendation, specific contraindication and harms, vaccine administration, and storage details were not mentioned. Specific indications were for everyone with coronary heart disease, unless contraindicated. A summary of the above vaccine recommendations can be found in the Supplementary Materials (Table S1).

3.2. Hypertensive Heart Disease

For hypertensive heart disease, 10 out of 76 guidelines mentioned vaccine recommendations. These guidelines were from Canada, United States, Europe, United Kingdom and Australia–New Zealand. All ten guidelines recommend the influenza vaccine. Eight out of the ten guidelines recommended the pneumococcal vaccine. One guideline recommended the COVID-19 vaccine.

This paragraph focuses on guidelines recommending the influenza vaccine. In terms of their strength of recommendation, two guidelines mentioned that it was recommended, while eight did not state the strength of recommendation. Additionally, specific indications from the ten guidelines include patients at high risk of developing heart failure and patients with heart failure. The vaccine is recommended for administration annually. Overall, specific contraindications and harms, vaccine administration, and storage details were not mentioned.

This paragraph focuses on guidelines recommending the pneumococcal vaccine. In terms of their strength of recommendation, two guidelines mentioned that it was recommended, while six did not state the strength of recommendation. Additionally, specific indications include patients at high risk of developing heart failure and patients with heart failure. Overall, specific contraindications and harms, vaccine administration, and storage details were not mentioned.

The strength of recommendation of the COVID-19 vaccine was not stated. It is indicated for patients with heart failure. Specific contraindications and harms, vaccine administration, and storage details were not mentioned. A summary of the above vaccine recommendations can be found in the Supplementary Materials (Table S2).

3.3. Atrial Fibrillation and Flutter

For atrial fibrillation and flutter, 1 out of 72 guidelines mentioned vaccine recommendations. That one guideline, from the United States, recommended the influenza and pneumococcal vaccines. The strength of recommendation of both is not stated. They are both indicated for patients with valvular heart disease. Overall, specific contraindications and harms, vaccine administration, and storage details were not mentioned. A summary of the above vaccine recommendations can be found in the Supplementary Materials (Table S3).

3.4. COPD

For COPD, 7 out of 39 guidelines mentioned vaccine recommendations. These guidelines were international, and from the United States, Canada, United Kingdom and Australia–New Zealand. All seven guidelines recommended the influenza vaccine. Only six out of seven guidelines recommended the pneumococcal vaccine. Only one guideline, from the Global Initiative for Chronic Obstructive Lung Disease (GOLD), recommended three additional vaccines: coronavirus disease 2019 (COVID-19); the tetanus, diphtheria, pertussis vaccine (Tdap); and zoster, alongside the two aforementioned vaccines.

This paragraph focuses on guidelines recommending the influenza vaccine. In terms of their strength of recommendation, two guidelines were strong, one weak, and four did not state the strength of recommendation. The specific indication for the influenza vaccine is for all patients with COPD. The vaccine is recommended for administration annually. Overall, specific contraindications and harms, vaccine administration, and storage details were not mentioned.

This paragraph focuses on guidelines recommending the pneumococcal vaccine. In terms of their strength of recommendation, two guidelines were weak, while four did not state this factor. The specific indications include COPD patients below 65 years old, and patients 65 years old and older; adults with COPD, especially those with specific comorbidities or undergoing certain treatments (e.g., chemotherapy); and varying recommendations, depending on smoking and vaccination history. The vaccine is recommended for administration at 50, 65 or at diagnosis of COPD, depending on the patient's smoking and vaccination history. This is followed by a second, subsequent revaccination. Overall, specific contraindications and harms, vaccine administration, and storage details were not mentioned.

For COVID-19, the tetanus, diphtheria, pertussis vaccine (Tdap), and zoster vaccines, the strength of recommendation was not stated in the guidelines. The specific indications include all patients with COPD, for the COVID-19 vaccine; those who were not vaccinated in adolescence, for the Tdap vaccine; and adults with COPD $\geq$ 50 years old, for the Zoster vaccine. For COVID-19, in terms of the timing of the vaccine, the guideline recommended following national guidelines. Overall, specific contraindications and harms, timing for the other two vaccines, vaccine administration, and storage details were not mentioned. A summary of the above vaccine recommendations can be found in the Supplementary Materials (Table S4).

3.5. Asthma

For asthma, 3 out of 45 guidelines mentioned vaccine recommendations. These guidelines were international, and from the United States and Australia–New Zealand. All three guidelines recommended the Influenza vaccine. Two out of three guidelines recommended the pneumococcal vaccine. Collectively, one other vaccine (the COVID 19 vaccine) was recommended, alongside the aforementioned vaccines.

Strength of recommendation for all vaccines was not stated in any of the three guidelines. Specific indications include patients with asthma; patients with severe asthma, defined as those who need frequent hospital visits and multiple medicines for asthma; all adults 65 years or more; patients with COPD; pregnant women; and any adult who wishes to avoid influenza. Contraindications to the influenza and pneumococcal vaccines are patients who are receiving high-dose oral steroid therapy. The influenza vaccine is recommended for administration annually.

Special considerations include that the influenza vaccine should not be given with the expectation that it will reduce either the frequency or severity of asthma exacerbations during the influenza season; that for patients who have documented histories of anaphylactic reactions after ingestion of egg protein and documented evidence of current allergic sensitization to eggs (skin testing or in vitro antigen-specific IgE antibody testing), the risk/benefit ratio of administering of influenza vaccine should be reviewed carefully; and that the first dose of biologic therapy and COVID-19 vaccine should not be given on the same day, to allow the adverse effects of either to be more easily distinguished. Overall, other specific contraindications and harms, vaccine administration, and storage details were not mentioned. A summary of the above vaccine recommendations can be found in the Supplementary Materials (Table S5).

3.6. Cirrhosis and Other Chronic Liver Diseases

For cirrhosis and other chronic liver diseases, 3 out of 45 guidelines mentioned vaccine recommendations. These guidelines were from the United States. All three guidelines recommended the hepatitis A and hepatitis B vaccines. Two out of three of the guidelines recommended the pneumococcal and influenza vaccines. Collectively, six other vaccines—Tdap, zoster, HPV, MMR, varicella, and COVID-19 vaccines—were recommended, alongside the previously mentioned vaccines.

Strength of recommendation for all vaccines was not stated in any of the three guidelines. Specific indications include patients with chronic liver disease and patients with alcoholic cirrhosis. Overall, specific contraindications and harms, vaccine administration, and storage details were not mentioned for all vaccines. A summary of the above vaccine recommendations can be found in the Supplementary Materials (Table S6).

3.7. Colon and Rectal Cancer

For colon and rectal cancer, 1 out of 46 guidelines mentioned vaccine recommendations. That 1 guideline, from India, recommended the influenza, hepatitis B, MMR, BCG and yellow fever vaccines. The strength of recommendation for all the vaccines was not stated. They are indicated for patients with colorectal cancer undergoing chemotherapy. Some contraindications include that the MMR, BCG and yellow fever vaccines should never be administered to immunocompromised patients, including those receiving chemotherapy, within 6 months of receiving chemotherapy. In terms of timing, the influenza vaccine should be given before chemotherapy, and the hepatitis B vaccine should be given at the end of the chemotherapy cycle. Overall, immunization should be postponed if a patient is suffering from an acute illness. Other vaccine administration information and storage details were not mentioned. A summary of the above vaccine recommendations can be found in the Supplementary Materials (Table S7).

3.8. Stomach Cancer

For stomach cancer, 2 out of 22 guidelines mentioned vaccine recommendations. These guidelines were from Canada and India. Both guidelines recommended the influenza, pneumococcal and haemophilus influenza type B (Hib) vaccine. Collectively, four other vaccines—Tdap, polio, varicella zoster, and meningococcal vaccines—were also recommended, on top of the previously mentioned vaccines.

This paragraph focuses on guidelines recommending the influenza, pneumococcal and haemophilus influenza type B (Hib) vaccine. The strength of recommendation was not stated. Additionally, specific indications include patients above 2 years old and patients with gastric lymphoma. The vaccine is recommended to be given 2 to 3 weeks before operation or initiation of anti-lymphoid cancer treatment, and given again 5 years later. Overall, specific contraindications and harms, vaccine administration, and storage details were not mentioned.

For the Tdap, polio, varicella zoster and meningococcal vaccines, the strength of recommendation was not stated. Specific indications include patients above 2 years old and patients with gastric lymphoma. It is recommended that vaccination is carried out once for the meningococcal vaccine and every 10 years for the Tdap and polio vaccines. The oral polio vaccine is contraindicated in patients with lymphoid cancer. Overall, other specific contraindications and harms, vaccine administration, and storage details were not mentioned. A summary of the above vaccine recommendations can be found in the Supplementary Materials (Table S8).

3.9. Diabetes

For diabetes, 7 out of 108 guidelines mentioned vaccine recommendations. These guidelines were international, and from Canada, the United States, United Kingdom, India, Australia–New Zealand and Japan. All seven guidelines recommended the influenza vaccine. Only six out of seven guidelines recommended the pneumococcal vaccine. Collectively, seven other vaccines—Tdap, hepatitis A, hepatitis B, herpes zoster, varicella, human papillomavirus (HPV), and measles, mumps, rubella (MMR)—were also recommended, alongside the two previously mentioned vaccines.

This paragraph focuses on guidelines recommending the influenza vaccine. In terms of their strength of recommendation, one guideline was very strong, one weak and five not stated. Additionally, specific indications from the seven guidelines include all older people with diabetes, and persons with diabetes who are 6 months old and older. One guideline indicated a contraindication of egg allergy, a recent history of Guillain–Barre syndrome within six weeks of a previous influenza vaccination, and febrile illness or any acute infection. The vaccine is recommended for administration annually. Overall, specific contraindications and harms, vaccine administration, and storage details were not mentioned.

This paragraph focuses on guidelines recommending the pneumococcal vaccine. In terms of their strength of recommendation, one guideline was weak, while five were not stated. Additionally, specific indications from the six guidelines include persons with diabetes aged 19 to 64 years, and people with diabetes 65 years and older or with an immunocompromising condition (e.g., end-stage renal disease). One guideline indicated a contraindication of hypersensitivity to the active substances, or to any of the excipients of the vaccine febrile illness, or any acute infection. The vaccine is recommended for administration at the time of diagnosis, with a second and third dose later on in life. Overall, specific contraindications and harms, vaccine administration, and storage details were not mentioned.

For the seven other vaccines—Tdap, hepatitis A, hepatitis B, herpes zoster, varicella, human papillomavirus (HPV), and measles, mumps, rubella (MMR)—the strength of recommendation for the guidelines varied from not stated to very strong. Specific indications include all unvaccinated patients with diabetes, for the hepatitis B vaccine; females and diabetic patients, for the HPV vaccine; and T2DM patients aged 70 to 79 years old, for the zoster vaccine. In terms of the timing of the vaccinations, the guidelines recommend that the hepatitis B vaccine be given at the diagnosis of diabetes; that Tdap be given every 10 years, following the completion of the primary series in routine childhood vaccination; that one or two doses of MMR vaccine be given 4 weeks apart; that two doses of Varicella vaccine be given 4 weeks apart; that the zoster vaccine be given once at 60 years old; that two doses of the hepatitis A vaccine be given 6 months apart; and that three doses of the HPV vaccine be given, up to the age of 26. Overall, specific contraindications and harms, vaccine administration, and storage details were not mentioned. A summary of the above vaccine recommendations can be found in the Supplementary Materials (Table S9).

3.10. CKD

For chronic kidney disease, 6 out of 46 guidelines mentioned vaccine recommendations. These guidelines were international, and from the United States, Australia–New Zealand and Japan. Five out of six of the guidelines recommended both the pneumococcal and influenza vaccine. Four out of six of the guidelines recommended the hepatitis B vaccine. Collectively, six other vaccines—Tdap, MMR, varicella, zoster/shingles, varicella, hepatitis A—were also recommended alongside the three previously mentioned vaccines.

This paragraph focuses on guidelines recommending the influenza vaccine. Their strength of recommendation ranged from a grade 1 strength of recommendation to a grade 2 level of recommendation (i.e., strong); some guidelines did not state a strength of recommendation. Additionally, specific indications from the six guidelines include adults with CKD and/or diabetes. Specific contraindications mentioned include giving a live attenuated influenza vaccine to CKD patients. The vaccine is recommended for administration annually. Overall, storage details were not mentioned.

This paragraph focuses on guidelines recommending the pneumococcal vaccine. Their strength of recommendation ranged from a grade 1 strength of recommendation to a grade 2 level of recommendation (i.e., strong); some guidelines did not state a strength of recommendation. Additionally, specific indications include adults aged $\geq$ 19 years with immunocompromising conditions (including those with chronic renal failure or nephrotic syndrome), functional or anatomic asplenia, cerebrospinal fluid (CSF) leaks, or cochlear implants; eGFR < 30 mL/min/1.73 m^2 (GFR categories G4–G5), and those at high risk of pneumococcal infection (e.g., nephrotic syndrome, diabetes, or those receiving immunosuppression); all adults aged 65 years and older; and adults at high risk aged 19 to 64 years. Revaccination is recommended within 5 years. Overall, specific contraindications and harms, vaccine administration, and storage details were not mentioned.

For the six other vaccines (Tdap, MMR, varicella, zoster/shingles, varicella, and hepatitis A), the strength of recommendation for the guidelines varied from a grade 1 strength of recommendation to strong. Specific indications include all susceptible chronic haemodialysis patients; pre-end-stage renal disease patients before they become dialysis dependent; and a history of HCV infection (whether NAT-positive or not). The hepatitis B vaccine requires booster doses, with a four-dose schedule (20 ug [1.0 mL doses]) administered in one or two injections. Overall, for the rest of the vaccines, specific contraindications and harms, vaccine administration, and storage details were not mentioned. A summary of the above vaccine recommendations can be found in the Supplementary Materials (Table S10).

3.11. Stroke

For stroke, 3 out of 62 guidelines mentioned vaccine recommendations. These guidelines were from the United States and Canada. All three guidelines recommended the influenza vaccine. The guidelines from the United States recommended that the vaccinations be taken annually.

In terms of their strength of recommendation, one guideline had level B evidence from randomized controlled trials, one was moderate, and one did not state a strength of recommendation. Only one guideline mentioned specific indications, which was in patients with pre-existing cardiovascular risk factors. Overall, specific contraindications and harms, vaccine administration, and storage details were not mentioned. A summary of the above vaccine recommendations can be found in the Supplementary Materials (Table S11).

4. Discussion

Detailed information about the vaccinations, including contraindications, route of administration, dosage, and storage, is mostly missing in the guidelines to be discussed in this section. This could be because many countries have national public health agencies that consolidate this information. These include the Centers for Disease Control and Prevention in the United States [16], and the Department of Health and Aged Care in Australia [17].

However, it should be noted that there is more information regarding the recommended influenza vaccine. This could be due to the fact that influenza vaccines are well established in terms of their development, mechanism of action, ingredients, and contraindications [18].

4.1. Discussion for Each Recommended Vaccine

4.1.1. Influenza Vaccine

In general, individuals with pre-existing diseases are at much higher risk of complications of influenza infections [19]. Influenza vaccines have been found to effectively decrease mortality due to influenza infections among elderly patients [20].

The following is a discussion about the rationale behind recommending the influenza vaccination to patients with the respective diseases.

Cardiac-Related Diseases

Influenza vaccination has evidently lowered the risk of cardiac and non-cardiac mortality in elderly patients with ischaemic heart disease [21,22]. The influenza vaccine also has evidently reduced the overall morbidity and mortality of diabetic patients and those with hypertensive heart disease [21,23,24].

Patients with underlying atrial fibrillation have a worse prognosis with influenza infection [25]. Influenza infection also increases the risk of haemorrhagic [26] and ischaemic stroke [27] in patients with atrial fibrillation. Therefore, the influenza vaccination is recommended for its protective effects.

Respiratory-Related Diseases

Influenza infection is known to be associated with COPD exacerbations, stroke, respiratory failure and pneumonia in patients with COPD [28]. Influenza vaccination can effectively reduce these events in COPD patients [29]. Similarly, in patients with asthma, influenza vaccination effectively lowers the rate of asthma exacerbations requiring Accident and Emergency department visits and/or hospitalisations, and influenza infections [30,31].

Gastrointestinal-Related Diseases

The efficacy of the influenza vaccine in patients with chronic liver disease is uncertain, due to the lack of research studies with a substantial sample size [32,33]. However, since influenza infection is known to cause the decompensation of liver cirrhosis and to increase mortality [34], many still recommend the influenza vaccination for patients with liver cirrhosis [35].

Cancer patients, such as those with colorectal and stomach cancer, especially those undergoing or planning to undergo chemotherapy, are recommended for vaccination against influenza. This is because influenza infection has been observed to delay chemotherapy [36], and results in severe complications in these immunocompromised patients [37]. As such, the influenza vaccine has been associated with lower mortality in cancer patients [38].

Other Diseases

In diabetic patients, influenza vaccination has been proven to reduce the overall morbidity and mortality related to influenza infection and cardiovascular events [39,40].

The influenza vaccine has been found to lower the morbidity of CKD patients from coronary heart disease [41], heart failure [42], and dementia [43], and even decreases the incidence of lung cancer in patients with CKD [44].

Studies have also revealed that influenza vaccination reduces the risk of all types of strokes, to a varying extent, regardless of the baseline risk of stroke [45], as influenza infection has been suggested to precipitate stroke [46].

Vaccine Administration Details

Most guidelines did not specify which type of influenza vaccine is recommended. For diabetes patients, specific types of influenza vaccines were recommended.

In guidelines written for diabetes, two guidelines specifically recommended the quadrivalent influenza vaccine, which has been available in the developed nations since 2012. Quadrivalent vaccines include an additional strain of influenza B virus compared to the trivalent vaccines. It has been estimated that the quadrivalent influenza vaccine has higher protective effects and higher cost-effectiveness than trivalent influenza vaccines [47,48].

Some guidelines for hypertensive heart disease and stomach cancer have specifically recommended that the elderly be vaccinated in fall, because studies have shown that the incidence of influenza infection peaks in winter [49].

For patients with colorectal cancer, it is recommended that influenza vaccination be offered before chemotherapy for optimal protection during chemotherapy treatment, as this treatment causes the patient to be in an immunocompromised state [50].

4.1.2. Pneumococcal Vaccine

Those with pre-existing medical conditions are immunocompromised and are at a higher risk of contracting pneumococcal pneumonia and having complications. Pneumococcus is also the most common cause of community-acquired pneumonia, demonstrating high incidence rates of invasive pneumococcal disease among adults above the age of 65 [51].

Cardiac-Related Diseases

In patients with cardiovascular disease with pneumococcal infections, the severity and risk of complications is higher. Hypertensive heart disease and atrial fibrillation, two of the cardiac conditions analyzed in this paper, are associated with a higher risk of stroke, cardiovascular disease (such as ischaemic heart disease), and even cardiovascular death [52,53]. Pneumococcal vaccinations have cardioprotective effects, reducing the risk of myocardial infarctions and decreasing morbidity and mortality.

Respiratory-Related Diseases

COPD patients are more susceptible to respiratory infections due to the impaired mucociliary clearance mechanisms and increased mucus production, which allow for increased bacterial and viral attachment. Use of pneumococcal vaccinations in COPD patients prevents exacerbations from respiratory tract infections, and is therefore recommended [54]. Guidelines for asthma were also analyzed in this study. Respiratory infections, especially community-acquired pneumonia, are one of the main causes of asthma exacerbations [55]. Use of the pneumococcal vaccination among asthma patients has been documented to have prompted a decrease in pneumococcal pneumonia-related hospitalizations [56].

Gastrointestinal-Related Diseases

Patients with severe liver disease have increased mortality and morbidity due to S. pneumoniae infections. In a study conducted among 45 unimmunised patients with end-stage liver disease who were vaccinated during liver transplantation evaluation, a significant response to the 23-valent pneumococcal vaccine was observed [57]. As such, the pneumococcal vaccine plays a critical role in the management of patients with chronic liver disease.

The pneumococcal vaccine is also recommended for patients with stomach cancer. Cancer patients are at higher risk of invasive pneumococcal disease (IPD) compared to the general population, with immunocompromised cancer patients contributing to 17–37% of all IPD cases [58].

The vaccine was also not recommended in the colorectal cancer guidelines. A study conducted in Taiwan identified 120,605 elderly patients with colorectal cancer, and explored the effectiveness of the pneumococcal vaccine in these patients. It found that the PPSV23 vaccine significantly reduced the rate of pneumonia hospitalization in elderly patients, and that the pneumonia-free survival rate was significantly higher in vaccinated patients compared to unvaccinated ones [59]. The lack of pneumococcal vaccine recommendations in the guidelines could be due to the fact that only one guideline even mentioned vaccines. A larger database of guidelines might yield better results.

Other Diseases

Patients with DM are at increased risk of acquiring pneumonia and invasive pneumococcal disease. They are also six times more likely to be hospitalized, and three times more likely to die from complications of influenza or pneumonia, than those in the general population [60]. The use of pneumococcal vaccinations significantly lowers the risk of morbidity and mortality.

The use of PPV-23 and PCV-13 is recommended in CKD patients. CKD patients have decreased B and CD4+ lymphocytes, and are at high risk of infections. Streptococcus pneumoniae is one of the main causes of community-acquired pneumonia in dialysis and kidney transplant patients [61].

The pneumococcal vaccine was not mentioned in any of the stroke guidelines. This could be due to the fact that many clinical studies report the vaccine having no effect on stroke risk. A study conducted by Kaiser Permanente concluded that the pneumococcal vaccine was not associated with reduced stroke risk. There were 5.30 stroke events per 1000 vaccinated person years, and 1.90 per 1000 unvaccinated person years [62]. These results were corroborated by another study based on data from the United Kingdom.

The pneumococcal vaccine was found to have no significant effect on stroke or transient ischaemic attack risk [63].

Vaccine Administration Details

Out of the 33 guidelines from different countries, among five diseases, only 9 guidelines provided information about the type of pneumococcal vaccine recommended. Five guidelines recommended both the PPV-23 and PCV-13 vaccines. Five guidelines recommended only the PPV-23 vaccine. One guideline from the American Association of Clinical Endocrinology recommended the PCV 15 and 20 vaccines as well.

In a study conducted among outpatients aged ≥ 65 years with chronic respiratory diseases in Shizuoka General Hospital, Japan, PPSV-23 was proven to be effective in preventing pneumococcal pneumonia among the elderly. Out of 320 patients vaccinated with PPSV-23, 1.88% developed pneumococcal pneumonia compared to 4.05% of the 3898 unvaccinated patients [64]. The effectiveness of the PPSV-23 vaccine explains why the majority of guidelines recommend the PPV-23 vaccine.

The combined use of PPSV-23 and PCV-13 is also highly recommended by several guidelines. A 2013 study involving 936 adults aged 70 years and older highlighted the limitations of using PPSV23 alone, and suggested that it might be more effective to administer PCV13 following initial vaccination with PPSV-23 [65]. The combination of PPSV-23 with PCV-13 has been documented to produce a superior immune response to PPSV-23 alone, which would allow for an overall reduction in the severity of pneumococcal pneumonia.

There are specific indications stated for chronic kidney disease. The pneumococcal vaccine is recommended for patients with eGFR < 30 mL/min/1.73 m^2 (GFR categories G4–G5) and those at high risk of pneumococcal infection (e.g., nephrotic syndrome, diabetes, or those receiving immunosuppression). As CKD progresses, hospitalization rates and the risk of pneumococcal infection increases. Therefore, patients with more severe CKD must be protected against pneumococcal pneumonia [66].

As for the timing of the vaccine, the PPSV-23 vaccine is recommended for administration at time of diagnosis, and a single revaccination within 5 years. The second dose of PPSV-23 is recommended to be at least 5 years apart from the first dose. If the PSV-13 vaccine is co-administered, it should be given first and followed at least 8 weeks later by the PPSV-23 vaccine. The recommended interval between the PPSV-23 and PCV-13 vaccine is extended up to 1 year or even 5 years in some guidelines. If the PCV-15 vaccine is used, PPSV-23 should be administered at least 12 months after.

For stomach cancer, there are specific considerations involved in the timing of the pneumococcal vaccine. It is recommended for administration at least 2 weeks prior to initiation of anti-lymphoid cancer treatment or splenectomy, and to be repeated 5 years later (this applies to both vaccines). However, according to a study conducted in three cancer centres in Korea from March 2016 to March 2018, administering the vaccine on day 1 of treatment in patients with gastric cancer is not inferior to administering the vaccine 2 weeks prior [67]. The paper attributed the known timing of vaccine administration 2 weeks prior to treatment to the lack of studies exploring the optimal timing required.

4.1.3. Hepatitis Vaccine

Gastrointestinal-Related Diseases

A possible reason that only four diseases had a recommendation for hepatitis vaccines could be the increased susceptibility to hepatitis infections of patients who have these diseases and are immunocompromised. Diabetes has emerged as a risk factor for increased complications in patients with acute viral hepatitis [68]. The hepatitis B virus is one of the major causes of chronic liver disease, and it can cause many extrahepatic complications and manifestations, including renal failure and various nephropathies [69]. However, it has been found that vaccination of haemodialysis patients with a combined hepatitis A and hepatitis B vaccine results in increased seroprotection against the hepatitis B virus, compared to the hepatitis B monovalent vaccine [70]. This explains the recommendation that both the hepatitis A and B vaccine be given together (instead of a hepatitis B monovalent vaccine) in patients with chronic kidney disease.

Acute hepatitis A and B against the the background of chronic liver disease are associated with more severe liver disease and a higher fatality rate, thus explaining why hepatitis A and B vaccinations are recommended for these patients [71]. A possible explanation for India being the only country to recommend hepatitis vaccines for colon and rectal cancer is that viral hepatitis is a major health challenge in India, and therefore the hepatitis vaccine is recommended for various diseases, and not just specifically colon and rectal cancer [72]. This can be observed in a guideline from India that also recommends the hepatitis A vaccine for patients with diabetes.

Cardiac-Related Diseases

Conversely, there are possible reasons as to why the other seven non-communicable diseases did not recommend the hepatitis vaccine. There is research that suggests that hepatitis vaccines do not have an impact on decreasing the risk of ischaemic heart disease [73]. Additionally, there is also a report that mentions HBV infection being associated with a lower risk of developing stroke; however, further research is required to confirm this [74].

Respiratory-Related Diseases and Other Diseases

For respiratory-related diseases and other diseases, at the time of writing, there are no papers that provide strong evidence for contraindication to hepatitis vaccines or any possible side effects in the elderly population.

Vaccine Administration Details

In terms of the details of hepatitis vaccine recommendations, all the guidelines for cirrhosis and other chronic liver diseases do not state any strength of recommendation or any details of vaccination, such as specific contraindication, harm, timing, route, dosage, and storage. For the other three diseases, there is some information provided, but these details are largely missing.

4.1.4. COVID-19 Vaccine

COVID-19 vaccination is recommended by some guidelines concerning COPD, diabetes, chronic liver disease and asthma, published in 2022. This is because patients with COPD, asthma, diabetes, and liver cirrhosis are at higher risk of severe COVID-19 infection [75–77].

4.1.5. Varicella Zoster Vaccine

Studies have shown that patients with COPD, diabetes, chronic kidney disease, chronic liver disease, and stomach cancer are at higher risk of zoster infection [78]. This is particularly pertinent for patients with diabetes and/or renal diseases, as they have a 1.8- to 8.4-fold higher risk of zoster infections than patients with other underlying diseases [79]. There is a much lower risk of zoster reactivation, relative to other infections such as pneumococcus and hepatitis infections, in patients with chronic liver disease [80]. Varicella zoster vaccination is recommended for gastric lymphoma patients. Although no reasoning can be found specifically for gastric lymphoma patients, varicella zoster vaccines have been found to be immunogenic in patients with solid organ tumours, with no significant safety concerns [81].

4.1.6. Tdap Vaccine

Although we vaccinate children with Tdap vaccines, the seroprevalence of diphtheria, tetanus, and pertussis is low in the elderly [82]. Tdap vaccines and their boosters are recommended in the guidelines written for COPD, diabetes, chronic kidney disease, chronic liver disease, and stomach cancer. A relatively significant percentage of elderly people with severe pertussis infections in United States from 2011–2015 were found to have underlying diseases such as diabetes and renal dysfunction, suggesting a correlation between these diseases and the development of severe pertussis infection [83]. COPD and asthma also evidently increase the risk of severe pertussis infection; pertussis infection, similarly, exacerbates asthma and COPD [84]. This explains why Tdap vaccines are recommended for patients with these diseases; surprisingly, Tdap vaccination is not recommended in the guidelines written for asthma. There is limited knowledge on the efficacy of Tdap in patients with chronic liver disease and/or cirrhosis [32]. On the other hand, poor efficacy in elderly has been reported for vaccinations against diphtheria and tetanus [85–87]. Limited studies have evaluated the efficacy of tetanus vaccine in protecting patients with stomach cancers from Clostridium tetani, and instead, more studies are exploring the therapeutic effects of tetanus toxoid in treating stomach cancers [88].

4.1.7. MMR and Varicella Vaccine

There are mixed opinions regarding re-vaccinating elderly people with these childhood vaccines for protection against vaccine-preventable diseases. Some believe that since the immunity of the elderly has waned over the years, re-vaccinating therefore protects them, but there is a lack of evidence for this [89]. With the recent resurgence of measles infections, there is the potential of the benefits outweighing the risks of re-vaccinating the elderly [90].

The MMR vaccine is also recommended for patients with colorectal cancer who are no longer immunocompromised. It has been found that the seroprevalence of measles and mumps antibodies is low among cancer patients, meaning their risk of measles and mumps infection is increased during community outbreaks [91].

4.1.8. HPV Vaccine

It has been found that patients with diabetes have more extensive infections and higher chances of the recurrence of genital warts, which are caused by HPV [92]. The HPV vaccine has been shown to be effective in reducing the incidence of genital warts secondary to HPV [93]. Limited studies have been carried out, showing mixed results concerning the efficacy of the HPV vaccine in patients with chronic liver disease [94].

4.1.9. Meningococcal Vaccine

Gastric lymphoma patients that are undergoing or have undergone splenectomy are at a higher risk of infection with encapsulated organisms, and as such, they should be vaccinated against these bacteria every five years [95].

4.1.10. Poliovirus Vaccine

It has been found that unless the patients with gastric lymphoma have a low antibody titre or have undergone hematopoietic stem cell transplantation, revaccination with the polio vaccine is not absolutely necessary [96].

4.1.11. Yellow Fever Vaccine

Yellow fever vaccination is recommended by one of the guidelines written for colon and rectal cancer. However, no clear evidence has been found to substantiate this recommendation.

4.2. Strengths and Limitations

This study has several strengths, including disease identification and focusing on the age group of 75 and above.

The diseases to be referenced were selected from the Global Burden of Disease Study 2019. This study collects data about "premature death and disability from more than 350 diseases and injuries in 195 countries". The large sample size and its extensiveness across different countries makes this study more reliable and representative of the world population. Composite indicators such as incidence, prevalence, mortality, years of life lost (YLLs), years lived with disability (YLDs), and disability-adjusted life years (DALYs) are used to assess disease burden. These allow for quantifiable measures to compare the burden of various diseases.

The focus on the older age group is a great strength of this study. Vaccination guidelines vary across different age groups, especially between children and adults. Each country follows its childhood immunization schedule based on the WHO position paper on routine immunizations for children. The vaccination guideline recommendations for children are more established and adhered to than those for adults. With decreasing vaccination compliance and increasing prevalence of diseases causing mortality and morbidity among older persons, narrowing the focus to older persons allows for this study to be more valuable in advising healthcare policies and practices.

One limitation of this study is the use of UpToDate to identify the guidelines to be examined. Although UpToDate is a resource accredited and recognized by experts and institutions around the world, there might be selection bias in the information and guidelines chosen to be listed on it. Only guidelines from selected countries, including the United States, Europe, Canada, United Kingdom, Australia–New Zealand, Japan, etc., were listed. This is not an accurate representation of the global population, and the compiled vaccination guidelines may not be applicable to every country.

Another limitation is the exclusion of non-English guidelines due to language issues. Many papers from Asian countries such as Japan were in the authors' native language, and the information in the guidelines could therefore not be examined. This resulted in guidelines applicable to the Asian population being less represented in this study.

4.3. Clinical Implications and Future Directions

The current guidelines for older adults do not commonly suggest vaccination for certain conditions. For those that do, the recommendations often lack information on potential negative effects and the proper administration of the vaccine.

This may be because there is not enough evidence to support the guidelines; the guidelines may also need to include vaccination guidance more frequently, and provide more specific information.

A possible strategy to combat this is to collaborate with international organizations, such as the World Health Organization (WHO) or the Centers for Disease Control and Prevention (CDC), to share information, insights, and best practices. International collaboration can help fill information gaps and ensure standardisation in vaccine guidelines worldwide. Additionally, more time and resources should be allocated to conducting thorough scientific research on vaccines. This will involve working with reputable sources, such as scientific journals, public health organizations, and regulatory bodies, to gather as much information as possible about a given vaccine's potential negative effects and proper administration.

As a result, our work aims to encourage further research into vaccination for older adults, and to push for guidelines to provide more comprehensive recommendations. However, it is important to study whether these guideline improvements would ultimately lead to better clinical outcomes.

5. Conclusions

To conclude, vaccines still remain the most cost-effective health intervention, especially amongst vulnerable populations such as the elderly. Through our thorough review of guidelines for vaccination in adults 75 years old and above, we have identified many gaps that could potentially be filled and looked into by our global health system. Filling said gaps may potentially improve elderly people's quality of life, and reduce their disability-adjusted life years. Future studies should now follow up and review these guidelines, as more guidance is being put in place, including non-English guidelines.

Supplementary Materials: The following supporting information can be downloaded at: https://www.mdpi.com/article/10.3390/vaccines11061076/s1, Table S1: Summary of vaccination recommendation for IHD; Table S2: Summary of vaccination recommendation for Hypertensive heart disease; Table S3: Summary of vaccination recommendation for Atrial Fibrillation and Flutter; Table S4: Summary of vaccination recommendation for COPD; Table S5: Summary of vaccination recommendation for Asthma; Table S6: Summary of vaccination recommendation for Cirrhosis and Other Chronic Liver Diseases; Table S7: Summary of vaccination recommendation for Colon and rectum cancer; Table S8: Summary of vaccination recommendation for Stomach Cancer; Table S9: Summary of vaccination recommendation for Diabetes; Table S10: Summary of vaccination recommendation for CKD; Table S11: Summary of vaccination recommendation for Stroke.

Author Contributions: Introduction, H.S.; Materials and Methods, Y.C.H.; Results, A.R.I.; Discussion, H.S., Y.C.H. and A.R.I.; Conclusion, A.R.I.; Supervision, K.C.S. All authors have read and agreed to the published version of the manuscript.

Funding: This scientific review paper received no external funding.

Institutional Review Board Statement: Not applicable.

Informed Consent Statement: Not applicable.

Data Availability Statement: The data used in this research paper were obtained from Uptodate, an online medical resource. Access to Uptodate's database and content is subject to their terms of use and subscription agreement. Researchers interested in accessing the data used in this study should refer to Uptodate's website (www.uptodate.com) for information on how to obtain access to their platform.

Conflicts of Interest: The authors declare no conflict of interest.

References

1. World Health Organization. Global Vaccine Action Plan. Decade of vaccine collaboration. *Vaccine* **2013**, *31* (Suppl. 2), B5–B31. [CrossRef] [PubMed]
2. Berg, R.L.; Cassells, J.S. The Second Fifty Years: Promoting Health and Preventing Disability. In *The Second Fifty Years: Promoting Health and Preventing Disability*; National Academies Press: Washington, DC, USA, 1992. [CrossRef]
3. Mouton, C.P.; Bazaldua, O.V.; Pierce, B.; Espino, D.V. Common infections in older adults. *Am. Fam. Physician* **2001**, *63*, 257–268. [PubMed]
4. Pan American Health Organization. Noncommunicable Diseases. Available online: https://www.paho.org/en/topics/noncommunicable-diseases#:~:text=The%20term%20NCDs%20refers%20to,diabetes%20and%20chronic%20lung%20illnesses (accessed on 28 May 2023).
5. World Health Organization. Noncommunicable Diseases: Mortality. Available online: https://www.who.int/data/gho/data/themes/topics/topic-details/GHO/ncd-mortality (accessed on 28 May 2023).
6. World Health Organization. Vaccines and Immunization. Available online: https://www.who.int/health-topics/vaccines-and-immunization#tab=tab_1 (accessed on 15 December 2022).
7. World Health Organization, Strategic Advisory Group of Experts (SAGE). *WHO Vaccine Position Papers*; World Health Organization: Geneva, Switzerland, 2020.
8. Abbafati, C.; Machado, D.B.; Cislaghi, B.; Salman, O.M.; Karanikolos, M.; McKee, M.; Abbas, K.M.; Brady, O.J.; Larson, H.J.; Trias-Llimós, S.; et al. Global burden of 369 diseases and injuries in 204 countries and territories, 1990-2019: A systematic analysis for the Global Burden of Disease Study 2019. *Lancet* **2020**, *396*, 1204–1222. [CrossRef]
9. Edson, R.S.; Beckman, T.J.; West, C.P.; Aronowitz, P.B.; Badgett, R.G.; Feldstein, D.A.; Henderson, M.C.; Kolars, J.C.; McDonald, F.S. A multi-institutional survey of internal medicine residents' learning habits. *Med. Teach.* **2010**, *32*, 773–775. [CrossRef]
10. Phua, J.; See, K.C.; Khalizah, H.J.; Low, S.P.; Lim, T.K. Utility of the electronic information resource UpToDate for clinical decision-making at bedside rounds. *Singap. Med. J.* **2012**, *53*, 116–120.
11. World Health Organization. *World Health Statistics 2023: Monitoring Health for the SDGs, Sustainable Development Goals*; World Health Organization: Geneva, Switzerland, 2023; p. 88.
12. Atkins, D.; Best, D.; Briss, P.A.; Eccles, M.; Falck-Ytter, Y.; Flottorp, S.; Guyatt, G.H.; Harbour, R.T.; Haugh, M.C.; Henry, D.; et al. Grading quality of evidence and strength of recommendations. *BMJ* **2004**, *328*, 1490. [CrossRef]
13. Steinberg, E.; Greenfield, S.; Wolman, D.M.; Mancher, M.; Graham, R. (Eds.) *Clinical Practice Guidelines We Can Trust*; The National Academies Press: Washington, DC, USA, 2011; p. 290. [CrossRef]
14. Karlsson, L.C.; Lewandowsky, S.; Antfolk, J. The association between vaccination confidence, vaccination behavior, and willingness to recommend vaccines among Finnish healthcare workers. *PLoS ONE* **2019**, *14*, e0224330. [CrossRef]

15. Yaqub, O.; Castle-Clarke, S.; Sevdalis, N.; Chataway, J. Attitudes to vaccination: A critical review. *Soc. Sci. Med.* **2014**, *112*, 1–11. [CrossRef]
16. Centers for Disease Control and Prevention (CDC). Vaccine Label Examples. Available online: https://www.cdc.gov/vaccines/hcp/admin/storage/guide/vaccine-storage-labels.pdf (accessed on 3 September 2022).
17. Australian Government Department of Health and Aged Care. *Australian Immunisation Handbook*; Australian Government Department of Health and Aged Care: Canberra, Australia, 2022. Available online: https://immunisationhandbook.health.gov.au/contents/about-the-handbook#citation (accessed on 19 December 2022).
18. Hannoun, C. The evolving history of influenza viruses and influenza vaccines. *Expert Rev. Vaccines* **2013**, *12*, 1085–1094. [CrossRef]
19. Nichol, K.L.; Wuorenma, J.; von Sternberg, T. Benefits of influenza vaccination for low-, intermediate-, and high-risk senior citizens. *Arch. Intern. Med.* **1998**, *158*, 1769–1776. [CrossRef]
20. Darvishian, M.; Gefenaite, G.; Turner, R.M.; Pechlivanoglou, P.; Van der Hoek, W.; Van den Heuvel, E.R.; Hak, E. After adjusting for bias in meta-analysis seasonal influenza vaccine remains effective in community-dwelling elderly. *J. Clin. Epidemiol.* **2014**, *67*, 734–744. [CrossRef]
21. Jaiswal, V.; Ang, S.P.; Yaqoob, S.; Ishak, A.; Chia, J.E.; Nasir, Y.M.; Anjum, Z.; Alraies, M.C.; Jaiswal, A.; Biswas, M. Cardioprotective effects of influenza vaccination among patients with established cardiovascular disease or at high cardiovascular risk: A systematic review and meta-analysis. *Eur. J. Prev. Cardiol.* **2022**, *29*, 1881–1892. [CrossRef]
22. Liu, I.F.; Huang, C.C.; Chan, W.L.; Huang, P.H.; Chung, C.M.; Lin, S.J.; Chen, J.W.; Leu, H.B. Effects of annual influenza vaccination on mortality and hospitalization in elderly patients with ischemic heart disease: A nationwide population-based study. *Prev. Med.* **2012**, *54*, 431–433. [CrossRef]
23. Vardeny, O.; Claggett, B.; Udell, J.A.; Packer, M.; Zile, M.; Rouleau, J.; Swedberg, K.; Desai, A.S.; Lefkowitz, M.; Shi, V.; et al. Influenza Vaccination in Patients With Chronic Heart Failure: The PARADIGM-HF Trial. *JACC Heart Fail.* **2016**, *4*, 152–158. [CrossRef]
24. Rodrigues, B.S.; David, C.; Costa, J.; Ferreira, J.J.; Pinto, F.J.; Caldeira, D. Influenza vaccination in patients with heart failure: A systematic review and meta-analysis of observational studies. *Heart* **2020**, *106*, 350–357. [CrossRef]
25. Bolton, A.; Thyagaturu, H.; Ashraf, M.; Carnahan, R.; Hodgson-Zingman, D. Effects of atrial fibrillation on outcomes of influenza hospitalization. *Int. J. Cardiol. Heart Vasc.* **2022**, *42*, 101106. [CrossRef]
26. Liu, J.C.; Wang, T.J.; Sung, L.C.; Kao, P.F.; Yang, T.Y.; Hao, W.R.; Chen, C.C.; Hsu, Y.P.; Wu, S.Y. Influenza vaccination reduces hemorrhagic stroke risk in patients with atrial fibrillation: A population-based cohort study. *Int. J. Cardiol.* **2017**, *232*, 315–323. [CrossRef]
27. Kao, P.F.; Liu, J.C.; Hsu, Y.P.; Sung, L.C.; Yang, T.Y.; Hao, W.R.; Lin, Y.C.; Wu, S.Y. Influenza vaccination might reduce the risk of ischemic stroke in patients with atrial fibrillation: A population-based cohort study. *Oncotarget* **2017**, *8*, 112697–112711. [CrossRef]
28. Liao, K.M.; Chen, Y.J.; Shen, C.W.; Ou, S.K.; Chen, C.Y. The Influence of Influenza Virus Infections in Patients with Chronic Obstructive Pulmonary Disease. *Int. J. Chron. Obstruct. Pulmon. Dis.* **2022**, *17*, 2253–2261. [CrossRef]
29. Li, Y.; Zhang, P.; An, Z.; Yue, C.; Wang, Y.; Liu, Y.; Yuan, X.; Ma, Y.; Li, K.; Yin, Z.; et al. Effectiveness of influenza and pneumococcal vaccines on chronic obstructive pulmonary disease exacerbations. *Respirology* **2022**, *27*, 844–853. [CrossRef]
30. Vasileiou, E.; Sheikh, A.; Butler, C.; El Ferkh, K.; von Wissmann, B.; McMenamin, J.; Ritchie, L.; Schwarze, J.; Papadopoulos, N.G.; Johnston, S.L.; et al. Effectiveness of Influenza Vaccines in Asthma: A Systematic Review and Meta-Analysis. *Clin. Infect. Dis.* **2017**, *65*, 1388–1395. [CrossRef] [PubMed]
31. Suarez-Varela, M.M.; Llopis, A.; Fernandez-Fabrellas, E.; Sanz, F.; Perez-Lozano, M.J.; Martin, V.; Astray, J.; Castilla, J.; Egurrola, M.; Force, L.; et al. Asthma and influenza vaccination in elderly hospitalized patients: Matched case-control study in Spain. *J. Asthma* **2018**, *55*, 391–401. [CrossRef] [PubMed]
32. Alukal, J.J.; Naqvi, H.A.; Thuluvath, P.J. Vaccination in Chronic Liver Disease: An Update. *J. Clin. Exp. Hepatol.* **2022**, *12*, 937–947. [CrossRef] [PubMed]
33. Harmala, S.; Parisinos, C.A.; Shallcross, L.; O'Brien, A.; Hayward, A. Effectiveness of influenza vaccines in adults with chronic liver disease: A systematic review and meta-analysis. *BMJ Open* **2019**, *9*, e031070. [CrossRef] [PubMed]
34. Zou, B.; Yeo, Y.H.; Jeong, D.; Sheen, E.; Park, H.; Nguyen, P.; Hsu, Y.C.; Garcia, G.; Nguyen, M.H. Higher mortality and hospital charges in patients with cirrhosis and acute respiratory illness: A population-based study. *Sci. Rep.* **2018**, *8*, 9969. [CrossRef] [PubMed]
35. Sobotka, L.A.; Mumtaz, K.; Hinton, A.; Conteh, L.F. The time to advocate for influenza vaccines in patients with cirrhosis is now. *Clin. Res. Hepatol. Gastroenterol.* **2022**, *46*, 101838. [CrossRef]
36. Taha, A.; Vinograd, I.; Sakhnini, A.; Eliakim-Raz, N.; Farbman, L.; Baslo, R.; Stemmer, S.M.; Gafter-Gvili, A.; Leibovici, L.; Paul, M. The association between infections and chemotherapy interruptions among cancer patients: Prospective cohort study. *J. Infect.* **2015**, *70*, 223–229. [CrossRef]
37. Kunisaki, K.M.; Janoff, E.N. Influenza in immunosuppressed populations: A review of infection frequency, morbidity, mortality, and vaccine responses. *Lancet Infect. Dis.* **2009**, *9*, 493–504. [CrossRef]
38. Vinograd, I.; Eliakim-Raz, N.; Farbman, L.; Baslo, R.; Taha, A.; Sakhnini, A.; Lador, A.; Stemmer, S.M.; Gafter-Gvili, A.; Leibovici, L.; et al. Clinical effectiveness of seasonal influenza vaccine among adult cancer patients. *Cancer* **2013**, *119*, 4028–4035. [CrossRef]
39. Wang, I.K.; Lin, C.L.; Chang, Y.C.; Lin, P.C.; Liang, C.C.; Liu, Y.L.; Chang, C.T.; Yen, T.H.; Huang, C.C.; Sung, F.C. Effectiveness of influenza vaccination in elderly diabetic patients: A retrospective cohort study. *Vaccine* **2013**, *31*, 718–724. [CrossRef]

40. Vamos, E.P.; Pape, U.J.; Curcin, V.; Harris, M.J.; Valabhji, J.; Majeed, A.; Millett, C. Effectiveness of the influenza vaccine in preventing admission to hospital and death in people with type 2 diabetes. *CMAJ* **2016**, *188*, E342–E351. [CrossRef]
41. Chen, C.I.; Kao, P.F.; Wu, M.Y.; Fang, Y.A.; Miser, J.S.; Liu, J.C.; Sung, L.C. Influenza Vaccination is Associated with Lower Risk of Acute Coronary Syndrome in Elderly Patients with Chronic Kidney Disease. *Medicine* **2016**, *95*, e2588. [CrossRef]
42. Fang, Y.A.; Chen, C.I.; Liu, J.C.; Sung, L.C. Influenza Vaccination Reduces Hospitalization for Heart Failure in Elderly Patients with Chronic Kidney Disease: A Population-Based Cohort Study. *Acta Cardiol. Sin.* **2016**, *32*, 290–298. [CrossRef]
43. Liu, J.C.; Hsu, Y.P.; Kao, P.F.; Hao, W.R.; Liu, S.H.; Lin, C.F.; Sung, L.C.; Wu, S.Y. Influenza Vaccination Reduces Dementia Risk in Chronic Kidney Disease Patients: A Population-Based Cohort Study. *Medicine* **2016**, *95*, e2868. [CrossRef]
44. Chen, C.C.; Wu, C.H.; Lin, C.H.; Chiu, C.C.; Yang, T.Y.; Lei, M.H.; Yeh, H.T.; Jian, W.; Fang, Y.A.; Hao, W.R.; et al. Influenza Vaccination and Risk of Lung Cancer in Patients with Chronic Kidney Disease: A Nationwide, Population-Based Cohort Study. *Cancers* **2022**, *14*, 2926. [CrossRef]
45. Holodinsky, J.K.; Zerna, C.; Malo, S. Association between influenza vaccination and risk of stroke in Alberta, Canada: A population-based study. *Lancet Public Health* **2022**, *7*, e914–e922. [CrossRef]
46. Siriwardena, A.N. Increasing evidence that influenza is a trigger for cardiovascular disease. *J. Infect. Dis.* **2012**, *206*, 1636–1638. [CrossRef]
47. De Boer, P.T.; Crepey, P.; Pitman, R.J.; Macabeo, B.; Chit, A.; Postma, M.J. Cost-Effectiveness of Quadrivalent versus Trivalent Influenza Vaccine in the United States. *Value Health* **2016**, *19*, 964–975. [CrossRef]
48. Thommes, E.W.; Ismaila, A.; Chit, A.; Meier, G.; Bauch, C.T. Cost-effectiveness evaluation of quadrivalent influenza vaccines for seasonal influenza prevention: A dynamic modeling study of Canada and the United Kingdom. *BMC Infect. Dis.* **2015**, *15*, 465. [CrossRef]
49. Reichert, T.A.; Simonsen, L.; Sharma, A.; Pardo, S.A.; Fedson, D.S.; Miller, M.A. Influenza and the winter increase in mortality in the United States, 1959°C1999. *Am. J. Epidemiol.* **2004**, *160*, 492–502. [CrossRef]
50. Vollaard, A.; Schreuder, I.; Slok-Raijmakers, L.; Opstelten, W.; Rimmelzwaan, G.; Gelderblom, H. Influenza vaccination in adult patients with solid tumours treated with chemotherapy. *Eur. J. Cancer* **2017**, *76*, 134–143. [CrossRef] [PubMed]
51. Berical, A.C.; Harris, D.; Dela Cruz, C.S.; Possick, J.D. Pneumococcal Vaccination Strategies. An Update and Perspective. *Ann. Am. Thorac. Soc.* **2016**, *13*, 933–944. [CrossRef] [PubMed]
52. Tackling, G.; Borhade, M.B. *Hypertensive Heart Disease*; StatPearls: Treasure Island, FL, USA, 2023.
53. Odutayo, A.; Wong, C.X.; Hsiao, A.J.; Hopewell, S.; Altman, D.G.; Emdin, C.A. Atrial fibrillation and risks of cardiovascular disease, renal disease, and death: Systematic review and meta-analysis. *BMJ* **2016**, *354*, i4482. [CrossRef] [PubMed]
54. Froes, F.; Roche, N.; Blasi, F. Pneumococcal vaccination and chronic respiratory diseases. *Int. J. Chronic Obstr. Pulm. Dis.* **2017**, *12*, 3457–3468. [CrossRef]
55. Czaicki, N.; Bigaj, J.; Zielonka, T.M. Pneumococcal Vaccine in Adult Asthma Patients. *Adv. Exp. Med. Biol.* **2021**, *1289*, 55–62. [CrossRef]
56. Laratta, C.R.; Williams, K.; Vethanayagam, D.; Ulanova, M.; Vliagoftis, H. A case series evaluating the serological response of adult asthma patients to the 23-valent pneumococcal polysaccharide vaccine. *Allergy Asthma Clin. Immunol.* **2017**, *13*, 27. [CrossRef]
57. McCashland, T.M.; Preheim, L.C.; Gentry-Nielsen, M.J. Pneumococcal Vaccine Response in Cirrhosis and Liver Transplantation. *J. Infect. Dis.* **2000**, *181*, 757–760. [CrossRef]
58. Lee, Y.J.; Huang, Y.-T.; Kim, S.J.; Kerpelev, M.; Gonzalez, V.; Kaltsas, A.; Papanicolaou, G. Trends in Invasive Pneumococcal Disease in Cancer Patients After the Introduction of 7-valent Pneumococcal Conjugate Vaccine: A 20-year Longitudinal Study at a Major Urban Cancer Center. *Clin. Infect. Dis.* **2017**, *66*, 244–253. [CrossRef]
59. Chiou, W.Y.; Hung, S.K.; Lin, H.Y.; Chen, L.C.; Hsu, F.C.; Tsai, S.J.; Yu, B.H.; Lee, M.S.; Li, C.Y. Effectiveness of 23-valent pneumococcal polysaccharide vaccine on elderly patients with colorectal cancer: A population-based propensity score-matched cohort study. *Medicine* **2019**, *98*, e18380. [CrossRef]
60. American Association of Diabetes Educators. *Vaccination Practices for Adults with Diabetes*; American Association of Diabetes Educators: Chicago, IL, USA, 2019.
61. Haddiya, I. Current Knowledge of Vaccinations in Chronic Kidney Disease Patients. *Int. J. Nephrol. Renovasc. Dis.* **2020**, *13*, 179–185. [CrossRef]
62. Tseng, H.F.; Slezak, J.M.; Quinn, V.P.; Sy, L.S.; Van den Eeden, S.K.; Jacobsen, S.J. Pneumococcal vaccination and risk of acute myocardial infarction and stroke in men. *JAMA* **2010**, *303*, 1699–1706. [CrossRef]
63. Siriwardena, A.N.; Asghar, Z.; Coupland, C.C. Influenza and pneumococcal vaccination and risk of stroke or transient ischaemic attack-matched case control study. *Vaccine* **2014**, *32*, 1354–1361. [CrossRef]
64. Masuda, T.; Nakatani, E.; Shirai, T.; Akamatsu, T.; Tamura, K.; Takahashi, S.; Tanaka, Y.; Watanabe, H.; Endo, Y.; Suzuki, T.; et al. Effectiveness of a 23-valent pneumococcal polysaccharide vaccine for the prevention of pneumococcal pneumonia in the elderly with chronic respiratory diseases: A case–control study of a single center. *BMC Pulm. Med.* **2021**, *21*, 123. [CrossRef]
65. Jackson, L.A.; Gurtman, A.; Rice, K.; Pauksens, K.; Greenberg, R.N.; Jones, T.R.; Scott, D.A.; Emini, E.A.; Gruber, W.C.; Schmoele-Thoma, B. Immunogenicity and safety of a 13-valent pneumococcal conjugate vaccine in adults 70 years of age and older previously vaccinated with 23-valent pneumococcal polysaccharide vaccine. *Vaccine* **2013**, *31*, 3585–3593. [CrossRef]

66. Portoles-Perez, J.; Marques-Vidas, M.; Picazo, J.J.; Gonzalez-Romo, F.; Garcia-Rojas, A.; Perez-Trallero, E.; Gil-Gregorio, P.; de la Camara, R.; Morato, M.L.; Rodriguez, A.; et al. Recommendations for vaccination against pneumococcus in kidney patients in Spain. *Nefrologia* **2014**, *34*, 545–551. [CrossRef]
67. Choi, W.; Kim, J.G.; Beom, S.H.; Hwang, J.E.; Shim, H.J.; Cho, S.H.; Shin, M.H.; Jung, S.H.; Chung, I.J.; Song, J.Y.; et al. Immunogenicity and Optimal Timing of 13-Valent Pneumococcal Conjugate Vaccination during Adjuvant Chemotherapy in Gastric and Colorectal Cancer: A Randomized Controlled Trial. *Cancer Res. Treat.* **2020**, *52*, 246–253. [CrossRef]
68. Singh, K.K.; Panda, S.K.; Shalimar; Acharya, S.K. Patients with Diabetes Mellitus are Prone to Develop Severe Hepatitis and Liver Failure due to Hepatitis Virus Infection. *J. Clin. Exp. Hepatol.* **2013**, *3*, 275–280. [CrossRef]
69. Noble, J.; Jouve, T.; Malvezzi, P.; Rostaing, L. Renal complications of liver diseases. *Expert Rev. Gastroenterol. Hepatol.* **2018**, *12*, 1135–1142. [CrossRef]
70. Tung, J.; Carlisle, E.; Smieja, M.; Kim, P.T.; Lee, C.H. A randomized clinical trial of immunization with combined hepatitis A and B versus hepatitis B alone for hepatitis B seroprotection in hemodialysis patients. *Am. J. Kidney Dis.* **2010**, *56*, 713–719. [CrossRef]
71. Keeffe, E.B. Hepatitis A and B superimposed on chronic liver disease: Vaccine-preventable diseases. *Trans. Am. Clin. Climatol. Assoc.* **2006**, *117*, 227–237; discussion 237–238.
72. Abraham, P. Viral hepatitis in India. *Clin. Lab. Med.* **2012**, *32*, 159–174. [CrossRef] [PubMed]
73. Sung, J.; Song, Y.M.; Choi, Y.H.; Ebrahim, S.; Davey Smith, G. Hepatitis B virus seropositivity and the risk of stroke and myocardial infarction. *Stroke* **2007**, *38*, 1436–1441. [CrossRef]
74. Wang, Y.; Xiong, J.; Chen, X.; Niu, M.; Chen, X.; Guan, Y.; Zheng, K.; Xu, K. Hepatitis B virus infection and decreased risk of stroke: A meta-analysis. *Oncotarget* **2017**, *8*, 59658–59665. [CrossRef] [PubMed]
75. Marjot, T.; Webb, G.J.; Barritt, A.S.t.; Moon, A.M.; Stamataki, Z.; Wong, V.W.; Barnes, E. COVID-19 and liver disease: Mechanistic and clinical perspectives. *Nat. Rev. Gastroenterol. Hepatol.* **2021**, *18*, 348–364. [CrossRef] [PubMed]
76. Huang, B.Z.; Chen, Z.; Sidell, M.A.; Eckel, S.P.; Martinez, M.P.; Lurmann, F.; Thomas, D.C.; Gilliland, F.D.; Xiang, A.H. Asthma Disease Status, COPD, and COVID-19 Severity in a Large Multiethnic Population. *J. Allergy Clin. Immunol. Pract.* **2021**, *9*, 3621–3628.e2. [CrossRef]
77. Orioli, L.; Hermans, M.P.; Thissen, J.P.; Maiter, D.; Vandeleene, B.; Yombi, J.C. COVID-19 in diabetic patients: Related risks and specifics of management. *Ann. Endocrinol.* **2020**, *81*, 101–109. [CrossRef]
78. Kawai, K.; Yawn, B.P. Risk Factors for Herpes Zoster: A Systematic Review and Meta-analysis. *Mayo Clin. Proc.* **2017**, *92*, 1806–1821. [CrossRef]
79. Hata, A.; Kuniyoshi, M.; Ohkusa, Y. Risk of Herpes zoster in patients with underlying diseases: A retrospective hospital-based cohort study. *Infection* **2011**, *39*, 537–544. [CrossRef]
80. Hudson, D.; Khanna, R.; Brahmania, M.; Qumosani, K.; Teriaky, A. A225 Vaccine-Preventable Diseases in Hospitalized Patients with End-Stage Liver Disease/Cirrhosis: A Nationwide Cohort Analysis. *J. Can. Assoc. Gastroenterol.* **2022**, *5*, 115–116. [CrossRef]
81. Vink, P.; Delgado Mingorance, I.; Maximiano Alonso, C.; Rubio-Viqueira, B.; Jung, K.H.; Rodriguez Moreno, J.F.; Grande, E.; Marrupe Gonzalez, D.; Lowndes, S.; Puente, J.; et al. Immunogenicity and safety of the adjuvanted recombinant zoster vaccine in patients with solid tumors, vaccinated before or during chemotherapy: A randomized trial. *Cancer* **2019**, *125*, 1301–1312. [CrossRef]
82. Tanriover, M.D.; Soyler, C.; Ascioglu, S.; Cankurtaran, M.; Unal, S. Low seroprevalance of diphtheria, tetanus and pertussis in ambulatory adult patients: The need for lifelong vaccination. *Eur. J. Intern. Med.* **2014**, *25*, 528–532. [CrossRef]
83. Mbayei, S.A.; Faulkner, A.; Miner, C.; Edge, K.; Cruz, V.; Pena, S.A.; Kudish, K.; Coleman, J.; Pradhan, E.; Thomas, S.; et al. Severe Pertussis Infections in the United States, 2011–2015. *Clin. Infect. Dis.* **2019**, *69*, 218–226. [CrossRef]
84. Macina, D.; Evans, K.E. Pertussis in Individuals with Co-morbidities: A Systematic Review. *Infect. Dis. Ther.* **2021**, *10*, 1141–1170. [CrossRef]
85. Grasse, M.; Meryk, A.; Schirmer, M.; Grubeck-Loebenstein, B.; Weinberger, B. Booster vaccination against tetanus and diphtheria: Insufficient protection against diphtheria in young and elderly adults. *Immun. Ageing* **2016**, *13*, 26. [CrossRef]
86. Weinberger, B. Adult vaccination against tetanus and diphtheria: The European perspective. *Clin. Exp. Immunol.* **2017**, *187*, 93–99. [CrossRef]
87. Weinberger, B.; Schirmer, M.; Matteucci Gothe, R.; Siebert, U.; Fuchs, D.; Grubeck-Loebenstein, B. Recall responses to tetanus and diphtheria vaccination are frequently insufficient in elderly persons. *PLoS ONE* **2013**, *8*, e82967. [CrossRef]
88. He, Q.; Gao, H.; Gao, M.; Qi, S.; Zhang, Y.; Wang, J. Anti-Gastrins Antiserum Combined with Lowered Dosage Cytotoxic Drugs to Inhibit the Growth of Human Gastric Cancer SGC7901 Cells in Nude Mice. *J. Cancer* **2015**, *6*, 448–456. [CrossRef]
89. Wong, L.P.; Wong, P.F.; AbuBakar, S. Vaccine hesitancy and the resurgence of vaccine preventable diseases: The way forward for Malaysia, a Southeast Asian country. *Hum. Vaccines Immunother.* **2020**, *16*, 1511–1520. [CrossRef]
90. Siani, A. Measles outbreaks in Italy: A paradigm of the re-emergence of vaccine-preventable diseases in developed countries. *Prev. Med.* **2019**, *121*, 99–104. [CrossRef]
91. Marquis, S.R.; Logue, J.K.; Chu, H.Y.; Loeffelholz, T.; Quinn, Z.Z.; Liu, C.; Stewart, F.M.; Carpenter, P.A.; Pergam, S.A.; Krantz, E.M. Seroprevalence of Measles and Mumps Antibodies Among Individuals With Cancer. *JAMA Netw. Open* **2021**, *4*, e2118508. [CrossRef]
92. Yong, M.; Parkinson, K.; Goenka, N.; O'Mahony, C. Diabetes and genital warts: An unhappy coalition. *Int. J. STD AIDS* **2010**, *21*, 457–459. [CrossRef] [PubMed]

93. Navarro-Illana, E.; Lopez-Lacort, M.; Navarro-Illana, P.; Vilata, J.J.; Diez-Domingo, J. Effectiveness of HPV vaccines against genital warts in women from Valencia, Spain. *Vaccine* **2017**, *35*, 3342–3346. [CrossRef] [PubMed]
94. Rhee, Y.; Sha, B.E.; Santos, C.A.Q. Optimizing Vaccination in Adult Patients With Liver Disease and Liver Transplantation. *Clin. Liver Dis.* **2020**, *15*, 63–68. [CrossRef] [PubMed]
95. Hammerquist, R.J.; Messerschmidt, K.A.; Pottebaum, A.A.; Hellwig, T.R. Vaccinations in asplenic adults. *Am. J. Health Syst. Pharm.* **2016**, *73*, e220–e228. [CrossRef]
96. Tsang, V. Vaccination recommendations for the hematology and oncology and post-stem cell transplant populations. *J. Adv. Pract. Oncol.* **2012**, *3*, 71–83. [CrossRef]

vaccines

MDPI

Systematic Review

Efficacy, Immunogenicity, and Safety of COVID-19 Vaccines in Patients with Autoimmune Diseases: A Systematic Review and Meta-Analysis

Alvina Widhani [1,2,*], Anshari Saifuddin Hasibuan [1], Retia Rismawati [1], Suzy Maria [1], Sukamto Koesnoe [1], Muhammad Ikrar Hermanadi [1], Youdiil Ophinni [3,4,5], Chika Yamada [5], Kuntjoro Harimurti [6], Aldean Nadhyia Laela Sari [1], Evy Yunihastuti [1] and Samsuridjal Djauzi [1]

1 Allergy and Clinical Immunology Division, Department of Internal Medicine, Faculty of Medicine, Universitas Indonesia, Dr. Cipto Mangunkusumo Hospital, Jakarta 10430, Indonesia; sorihsb@gmail.com (A.S.H.); retiarismawati@gmail.com (R.R.); suzyduri@gmail.com (S.M.); sukamtokoesnoe@gmail.com (S.K.); ikrar.hermanadi@gmail.com (M.I.H.); aldeannadhyia@gmail.com (A.N.L.S.); evy.yunihastuti@gmail.com (E.Y.); samsuridjal@yahoo.com (S.D.)
2 Department of Internal Medicine, Universitas Indonesia Hospital, Depok 16424, Indonesia
3 Division of Clinical Virology, Center for Infectious Diseases, Graduate School of Medicine, Kobe University, Kobe 650-0017, Japan; yophinni@panda.kobe-u.ac.jp
4 Department of Host Defense, Immunology Frontier Research Center, Osaka University, Osaka 565-0871, Japan
5 Center for Southeast Asian Studies, Kyoto University, Kyoto 606-8304, Japan; chika128@cseas.kyoto-u.ac.jp
6 Geriatric Division, Department of Internal Medicine, Faculty of Medicine, Universitas Indonesia, Dr. Cipto Mangunkusumo Hospital, Jakarta 10430, Indonesia; kuntjoro.harimurti@gmail.com
* Correspondence: alvina.widhani@gmail.com

Citation: Widhani, A.; Hasibuan, A.S.; Rismawati, R.; Maria, S.; Koesnoe, S.; Hermanadi, M.I.; Ophinni, Y.; Yamada, C.; Harimurti, K.; Sari, A.N.L.; et al. Efficacy, Immunogenicity, and Safety of COVID-19 Vaccines in Patients with Autoimmune Diseases: A Systematic Review and Meta-Analysis. *Vaccines* **2023**, *11*, 1456. https://doi.org/10.3390/vaccines11091456

Academic Editor: Kay Choong See

Received: 15 July 2023
Revised: 30 August 2023
Accepted: 1 September 2023
Published: 4 September 2023

Abstract: Patients with autoimmune diseases are among the susceptible groups to COVID-19 infection because of the complexity of their conditions and the side effects of the immunosuppressive drugs used to treat them. They might show impaired immunogenicity to COVID-19 vaccines and have a higher risk of developing COVID-19. Using a systematic review and meta-analysis, this research sought to summarize the evidence on COVID-19 vaccine efficacy, immunogenicity, and safety in patients with autoimmune diseases following predefined eligibility criteria. Research articles were obtained from an initial search up to 26 September 2022 from PubMed, Embase, EBSCOhost, ProQuest, MedRxiv, bioRxiv, SSRN, EuroPMC, and the Cochrane Center of Randomized Controlled Trials (CCRCT). Of 76 eligible studies obtained, 29, 54, and 38 studies were included in systematic reviews of efficacy, immunogenicity, and safety, respectively, and 6, 18, and 4 studies were included in meta-analyses for efficacy, immunogenicity, and safety, respectively. From the meta-analyses, patients with autoimmune diseases showed more frequent breakthrough COVID-19 infections and lower total antibody (TAb) titers, IgG seroconversion, and neutralizing antibodies after inactivated COVID-19 vaccination compared with healthy controls. They also had more local and systemic adverse events after the first dose of inactivated vaccination compared with healthy controls. After COVID-19 mRNA vaccination, patients with autoimmune diseases had lower TAb titers and IgG seroconversion compared with healthy controls.

Keywords: autoimmune; efficacy; immunogenicity; safety; vaccine; COVID-19

1. Introduction

As of 26 December 2022, there were more than 651 million cases of COVID and more than 6 million deaths reported worldwide [1]. It is important to understand that certain groups in the population are higher-risk groups who are more susceptible to severe COVID-19 infection. These groups consist of people who have comorbidities, such as cancer, chronic kidney disease, underlying lung disorders, diabetes, dementia, cardiac issues, HIV, other immunocompromised conditions, neurological diseases, and pregnancy [2].

One of these susceptible groups is people with autoimmune diseases because of the complexity of these conditions and the mechanisms underlying the therapeutic effects of the drugs used to treat them. Medications play a pivotal role in significantly improving the disease course and outcomes of autoimmune patients. However, the primary disadvantage of these medications is the immunosuppressive effect they have, which can enhance the risk of infections. Therefore, there is an emerging demand to prioritize COVID-19 vaccination for people with autoimmune conditions, as this prevents severe disease outcomes [3,4].

Vaccination is an effort to suppress the case numbers and severity of COVID-19 infections [5,6]. It has been established that vaccines can induce humoral and/or cellular immune responses to build protection against various infectious diseases, which is an ability also known as immunogenicity [5,7]. Not only does COVID-19 vaccination protect healthy individuals from getting infected, it also prevents those who are infected from getting severely ill, or even dying, from COVID-19 [5–7]. As of 26 December 2022, 13 billion doses of COVID 19 vaccine had been administered worldwide [1].

An additional cause of concern is that patients with systemic autoimmune diseases might show impaired immunogenicity to COVID-19 vaccines. These patients can have a higher risk of developing COVID-19 [8]. Besides the issue of decreased vaccine efficacy due to the use of immunosuppressive drugs, the safety of the COVID-19 vaccine is also a concern among these patients [9,10]. Certain vaccine antigens and their adjuvants, such as aluminum salts (alum), have been claimed to induce autoimmunity in numerous studies. Adjuvants are usually needed in inactivated and recombinant protein vaccines to boost the immunogenicity induced by the antigen [4]. Patients with autoimmune diseases are more susceptible to vaccination-induced autoimmune/autoinflammatory syndrome induced by adjuvants (ASIA) [11]. SARS-CoV-2 amino acid sequences cross-react with human cell sequences [12]. The antibody to the S1 spike protein of SARS-CoV-2 has a high affinity for transglutaminase 3 protein, transglutaminase 2 protein, anti-extractable nuclear antigen, nuclear antigen, and myelin basic protein [13]. Despite the evidence, this claim should be interpreted cautiously, as the temporal relationship between the vaccine and autoimmune events is still unclear [4]. There is also evidence that non-live vaccines, including those for influenza and pneumococcal virus, do not cause exacerbation of previously diagnosed autoimmune conditions [3,6].

In the third-phase clinical trial of ChAdOx1 nCoV-19 (AstraZeneca), a simian adenovirus-vectored vaccine, there was one case of transverse myelitis reported 14 days after vaccination [14]. A cohort study from the health registry in Denmark and Norway showed an increase in venous thromboembolism cases, including cerebral venous thrombosis, 28 days after ChAdOx1 nCoV-19, and a slight increase in thrombocytopenia and bleeding cases [15]. Another study reported 39 patients with thrombocytopenia and thrombosis 5–24 days after vaccination with ChAdOx1 nCoV-19. These patients were diagnosed with vaccine-induced thrombotic thrombocytopenia (VITT) or thrombosis with thrombocytopenia syndrome (TTS), which were suspected to be caused by platelets activating antibodies to platelet factor 4 [16–18]. Brill et al. reported autoimmune hepatitis 6 days after administration of the Pfizer–BioNTech COVID-19 vaccine in a 35-year-old woman. This case report could not conclude whether this was a causal relationship or only coincidence [19]. There were also reports of thrombocytopenia post the mRNA vaccine, which were diagnosed as secondary immune thrombocytopenia (ITP), but again, it could not be determined whether this was a coincidence or vaccine-induced ITP [20].

Despite COVID-19 vaccination being recommended, the efficacy, immunogenicity, and safety of COVID-19 vaccination in people with autoimmune diseases have not been discussed much. In addition, patients with autoimmune conditions and/or people taking immunosuppressants were excluded from clinical trials of approved COVID-19 vaccines [4,21]. Therefore, this systematic review aims to summarize the evidence on COVID-19 vaccine efficacy, immunogenicity, and safety in autoimmune patients.

2. Materials and Methods

The protocol for this study has been registered in PROSPERO with the registration number CRD42022337621. This study was conducted in accordance with the Preferred Reporting Items of the Systematic Review and Meta-Analysis (PRISMA) checklist [22].

2.1. Eligibility Criteria

The specific inclusion criteria for the systematic review and meta-analysis were as follows: (1) all randomized controlled trials (RCTs), non-randomized studies of interventions, cohort studies, case–control studies, and cross-sectional studies; (2) studies with autoimmune patients as the population (with the autoimmune condition existing prior to the intervention); (3) COVID-19 vaccination as the intervention; (4) efficacy, immunogenicity or safety as outcomes; and (5) publication in English. The exclusion criteria were as follows: (1) full text or data that cannot be accessed even though the corresponding author has been contacted.

2.2. Information Sources and Search Strategy

We included all articles on patients with autoimmune diseases published in English from 2020 to 2022. Electronic databases were searched using PubMed, Embase, EBSCOhost, ProQuest, MedRxiv, bioRxiv, SSRN, EuroPMC, and the Cochrane Center of Randomized Controlled Trials (CCRCT) from 6–26 September 2022 for studies evaluating the response to SARS-CoV-2 vaccines using a combination of keywords and medical subject headings. The keywords utilized were "autoimmune"; "vaccine" or "immunization" or "vaccination"; "COVID-19"; "efficacy"; "immunogenicity"; and "safety" or "adverse event" or "adverse effect", along with their synonyms and related terms incorporated by the appropriate Boolean operators. The detailed search strategy for articles is available in the Supplementary Materials (Table S1).

2.3. Data Extraction

Records were checked for duplicates using Zotero 6.0.19. Two independent researchers screened the literature search and assessed each study for inclusion by reading titles, abstracts, and full texts. Different opinions during data extraction were resolved by discussion and the inclusion of a study was decided by the two researchers. Relevant data were obtained from each eligible study by using an extraction sheet, which was prepared and approved by all the reviewers by reaching a consensus after screening for the eligible studies. Relevant data that were collected included study characteristics (authors, year, country, research setting, study design, study duration, sample size); participant characteristics (autoimmune diagnosis, age, sex, comorbidities); intervention (COVID-19 vaccine platform) and comparison; and outcomes (efficacy, immunogenicity, safety). Two independent researchers collected the data from each research article. The corresponding authors were contacted to obtain any information that was not explicitly available.

2.4. Outcome Measures

All studies describing the efficacy, immunogenicity, or safety of the COVID-19 vaccine in autoimmune patients were evaluated. The main outcomes were (1) breakthrough COVID-19 events, severity of infection, hospitalization, and mortality as markers of efficacy; (2) neutralizing antibodies, antibody titers, and seroconversion as markers of immunogenicity; and (3) flares or autoimmune relapses, local reactions, systemic reactions, and other adverse events as markers of safety.

The pooled efficacy, immunogenicity, and safety data after primary or booster doses of COVID-19 vaccine were evaluated. Efficacy was measured by the number of COVID-19 breakthrough infections, severity of COVID-19 infections, and hospitalizations and mortality related to COVID-19 infection. A COVID-19 breakthrough infection was defined as an infection after receiving the vaccination. Severity was defined by one of three levels of COVID-19 infection after vaccination: mild, moderate, or severe. Hospitalization was defined as the number of people who were taken to hospital as a result of COVID-19 infection. Mortality was defined as the number of people who died as a result of COVID-19 infection. Immunogenicity was defined as the ability of COVID-19 vaccines to stimulate an immune response, which was measured by the proportions of subjects with seroconversion (based on total IgG, as measured by ELISA) and with neutralizing antibodies (based on a plaque reduction neutralization test (PRNT) or surrogate virus neutralization test (sVNT), total IgG antibody titers (following WHO guidelines on translating results from different ELISA manufacturers into standardized binding antibody units (BAU)/mL) [23], and neutralizing activity (based on PRNT or sVNT, calculated as (1-OD value of sample/OD value of control) $\times$ 100%). Antibody titers were log-transformed prior to standardized mean difference (SMD) calculation. Where applicable, $PRNT_{50}$ titer was correlated with sVNT inhibition capacity [24], mean and standard deviation (SD) were estimated from median and interquartile range (IQR) [25], SDs were estimated from 95% confidence interval, and means and SDs were aggregated from multiple subgroups. Safety was measured by the number of autoimmune relapses, local symptoms (pain, erythema, bruising, etc.), systemic symptoms (fever, joint pain, flu like symptoms, fatigue, headache, muscle pain), and other adverse events occurring after receipt of a dose of COVID-19 vaccine.

2.5. Risk of Bias Evaluation

Risk-of-bias and quality-of-study evaluations were carried out by two independent researchers. The Risk of Bias (RoB) and Risk of Bias in Nonrandomized Studies of Interventions (ROBINS-I) tools were used for randomized controlled trials (RCT) and nonrandomized studies of interventions, respectively [26]. Cross-sectional and case-series studies were assessed using the Newcastle–Ottawa Quality Assessment Scales and The National Institutes of Health (NIH) quality assessment tool, respectively [27,28]. The certainty of evidence for the primary outcomes was evaluated using the Grades of Recommendation, Assessment, Development, and Evaluation (GRADE) system in eight domains: risk of bias, inconsistency, indirectness, imprecision, publication bias, large effects, plausible confounding, and dose–response gradient [26].

2.6. Data Synthesis

All outcomes were analyzed using Microsoft Excel and RevMan version 5.4 issued by Cochrane. Outcomes were reported as risk ratios for categorical data and standardized mean differences for numerical data, each with a confidence interval. Risk ratio was used to compare the risks of outcomes measured among patients with autoimmune diseases to healthy controls, while standardized mean difference was used to assess and pool continuous data, which was measured in a variety of ways. For analyzing continuous data conversion, guidelines from the Cochrane book were applied [28]. Heterogeneity was assessed using Higgins I^2 and considered significant at $I^2 > 60\%$ [28]. For significantly heterogeneous data, subgroup analysis was performed. Fixed-effects models were used for data with no substantial heterogeneity or which was considered homogeneous, whereas random effect models were used when there was significant heterogeneity. Data was displayed as a forest plot for meta-analysis.

3. Results

Our search retrieved 1054 records, of which 833 were duplicates and were excluded. The titles and abstracts of the remaining 221 published articles were screened, and 188 were assessed for eligibility via full-text evaluation. One hundred and twelve records did not meet the inclusion criteria after this full-text review, and were excluded. As a result, 76 full-text articles were selected for systematic review. Subsequently, 20 full-text articles were selected for meta-analysis, with 6, 18, and 4 articles included for efficacy, immunogenicity, and safety, respectively. The study flow chart is presented in Figure 1.

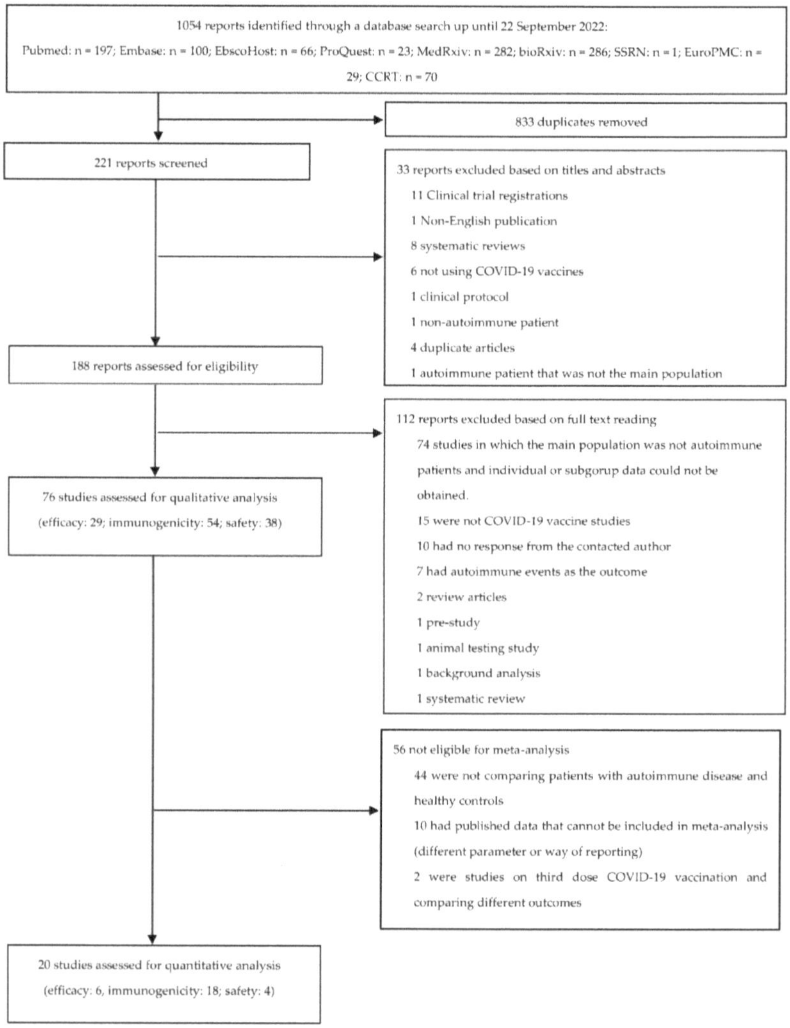

Figure 1. PRISMA flow chart.

3.1. Study Characteristics

Seventy-six studies were included in the qualitative analysis (Table S2). Ten studies were conducted in Israel [29–38], one study in Denmark [39], eight studies in Italy [8,40–46], three studies in the USA [47–49], nine studies in Germany [50–58], one study in New Zealand [59], three studies in Austria [60–62], four studies in Spain [63–66], one study in Japan [67], one study in France [68], one study in Romania [69], one study in Peru [70], one study in Canada [71], six studies in Brazil [72–77], two studies in China [78,79], three studies in Thailand [80–82], one study in Chile [83], five studies in India [84–88], one study in Greece [89], one study in Turkey [90], four studies in the UK [91–94], one study in Korea [95], one study in Taiwan [96], two studies in Netherlands [97,98], one study in Switzerland [85], one study in Hungary [99], and one study in the USA and UK [100]. The types of investigated studies encompassed single-blinded [73,96], observer-blinded randomized [31], and non-randomized [8,29,30,32–72,74–95,97–103] studies. A total of 160,447 participants were involved. All studies concerned adult participants (the majority of participants were >18 years of age), and only one study also involved pediatric participants [91]. Sixty-six studies included participants who had only had a primary dose vaccine [8,29,30,35–39,41,43–80,82–93,97–103], whereas in ten studies participants had had a booster dose vaccine [31–34,40,42,81,94–96].

The studies in our qualitative analysis were divided into six categories based on the type of vaccine: studies on mRNA vaccines including Pfizer/BioNTech (BNT162b2) and Moderna (mRNA-1273); studies on inactivated virus vaccines including CoronaVac, Covaxin (BBV152), and Sinopharm (BBIBP-CorV); studies on adenovirus vector vaccines including Vaxzevria (ChAdOx1), Janssen (Ad26.COV2.S), Sputnik V (Gam-COVID-Vac), and AstraZeneca (AZD1222); studies on mRNA vaccines and adenovirus vector vaccines; studies on inactivated virus and adenovirus vector vaccines; and studies on mRNA vaccines, inactivated virus vaccines, and adenovirus vector vaccines.

In terms of autoimmune diagnosis, studies included adult-onset Still's disease [57,61,89,95], antiphospholipid syndrome [33,47,69,72,74,75,88,89,94,96,102], autoimmune encephalitis [40,52,60], autoimmune hepatitis [44,56,69,90,97,101], autoimmune thyroid [69,96], IgG-4-related diseases [47,69,92,94], interstitial lung disease and systemic autoimmune disease/immune pulmonary disease [49,69], inflammatory bowel disease [47,50,62,69,71,91,97], inflammatory myopathies/systemic autoimmune myopathy [33,35,36,61,63,69,72,74–77,101], immune-mediated thrombocytopenic purpura/immune-mediated thrombotic thrombocytopenic purpura (ITP/iTTP) [32,42,96], juvenile idiopathic arthritis [57,83,88,89,94,98], mixed/undifferentiated connective tissue disease/connective tissue disease [44,47,50,55,57,61,69,86–89,94,98,103], multiple sclerosis [29–31,34,47,50,52–54,60,64,67,69,97,98], myasthenia gravis syndrome [52,60,69,97], neuromyelitis optica spectrum disorder [47,52,60,66,97], primary biliary cholangitis [56,69,97,101], psoriasis [50,69,71,80,97], psoriatic arthritis [33,35,50,57,69,71,72,74,75,83,88,94,98,99], rheumatoid arthritis (RA) [33,35,36,39,43,44,47,50,57,58,63,65,72–75,78,82–84,86–89,93–99,101,103], systemic lupus erythematosus (SLE) [33,35,36,39,44,45,47,50,57,61,65,66,68–70,72,74,75,77,78,81,82,84,86–89,92–99,101–103], sarcoidosis [50,61,88,94,98], spondiloarthritis/spondyloarthropathy [33,35,44,47,49,50,57,69,71,72,74,75,84,86–89,94–99,101,103], sclerosing cholangitis [56,97], Sjogren syndrome/sicca syndrome [33,47,61,65,69,71,74,75,78,88,89,94,96,98,99], systemic sclerosis [33,44,61,63,65,69,72,74,75,86–89,94,98,99,101,103], and vasculitides/vasculitis [33,35,36,44,47,50,55,57,58,61,66,69,72,74,75,86–89,92–95,97–99,103].

Autoimmune medications given to the patients included alemtuzumab [29,34,48,66], abatacept [8,33,35,39,47,57,58,63,72–75,94,98], anti-CD20/-B cell depleting therapy [8,29,32–35,37,39–41,43,45,47–49,52–55,57,60,62–66,72,74,75,87,89,92,94,96–98,103], antimalarials including hydroxychloroquine (HCQ) and chloroquine [8,37,39,41,44,45,47,50,55,61,63,65,68–70,72,75,77,81–84,87–90,93–96,98,99,103], apremilast [94,103], azathioprine [8,33,39–41,44,45,47,50,52,56–58,60,63,65,68–72,74–76,80–82,84,85,87–90,92,94,95,98,99,101,103], belimumab [8,39,41,44,45,47,50,57,61,65,68,72,74,75,77,89,92,94,95,99], calcineurin inhibitor [33,41,77,95], caplacizumab [42], certolizumab [50,83,101], cladibrine [29,30,34,48,52,66], colchicine [33,89,94,103], corticosteroids [32,33,35–37,39–42,44,45,47,49,50,52,55–58,60,62–64,68–70,72,73,75–77,80–84,86–94,96,98–101,103], cyclophosphamide (CYP) [45,62–64,72,74–77,87,89,92,95], cyclosporine (CYC) [45,72,74–76,80–83,85,88,89,94,98], denosumab [94], DMF [29,34,48,52,53,66], eculizumab [94], everolimus [85], fampridine [98], fingolimod [29–31,34,47,52,53,64,98], glatiramer acetate (GA) [29,34,48,52,53,64,98], ibrutinib [47,100], iguratimod [88], IL-1 inhibitor [89,94], IL-6 inhibitor [8,33,35,37,39,40,44,45,47,50,57,60,61,63,65,72,74,75,83,85,89,94,95,99,101], IL-17 inhibitor [33,35,44,51,57,71,72,74,75,80,83,89,94,98,99,101], IL-12/23 inhibitor [33,47,71,89,99], IL-23 inhibitor [47,71,99], β interferons [29,34,48,52,54,66,98], intravenous immunoglobulin (IVIG) [29,32,34,36,52,71,100], Janus kinase (JAK) inhibitor [8,33,35,39,57,58,62,88,89,94,95,99,101], leflunomide [8,36,39,45,47,50,58,63,69,70,72–77,82–84,87–89,93–95,98,99,103], lenalidomide [87,99], mepolizumab [99], methotrexate [8,33,35,36,39,41,43,44,47,57,58,61,63,65,68–77,80,82–84,87–90,92–99,101,103], mycophenolate mofetil [8,33,35–37,39–41,43–45,47,49,56,57,60,61,63,65,68–70,72,74–77,80–85,87–90,92,94,95,97,100,101,103], natalizumab [29,34,48,52,53,66,98], nintedanib [49], ocrelizumab (OCR) [29,30,34,48,52,60], ofatumumab [48], olumiant [61], omalizumab [80], pembrolizumab [99], plasmapheresis (PLEX) [42,52,64], sphingosine-1-phosphate receptor modulators (S1PRM) [48,66,97], salazopyrin [39], sulfasalazine [8,44,47,51,58,63,69,72,75,83,84, 87,93–98,98,103], tacrolimus [45,61,72,74,75,81–83,85,87,92,94,103], teriflunomide [29,34,48, 52,53,66], thalidomide [89,94], tumor necrosis factor alpha inhibitor (TNFi) [8,33,35,39,44,45, 47,50,51,57,58,62,63,65,71,72,74,83,87,89,91,94,95,97–99,101,103], tofacitinib [37,47,72,74,75, 87,103], upadacitinib [37,47], ustekinumab [50,72,74,75,98], and vedolizumab [47,50,91,92].

3.2. *Quality of Assessment*

Graphical representation of the studies' quality is illustrated in the Supplementary Materials (Figure S1A–D). Risks of bias in the three RCTs were low; twenty-one non-randomized studies were low-risk, thirty-seven non-randomized studies were moderate-risk, and seven non-randomized studies had serious risk; four case-series studies were defined as good; and four cross-sectional studies were considered fair.

3.3. *Qualitative Analysis*

3.3.1. Efficacy

In the mRNA vaccine studies group, efficacy after primary vaccination was reported as breakthrough COVID-19 infections [29,35,41,49,63,101], hospitalizations [49,63], and deaths [29,35,49,63]. Efficacy after booster vaccination was also reported as breakthrough COVID-19 infections [33,34] and hospitalizations and deaths [33]. In the inactivated virus vaccine studies, six studies reported breakthrough COVID-19 infections after primary vaccination as outcomes [72,75–77,80,83], two studies reported hospitalizations [76,77], and only one study reported death [77]. One study on adenovirus vector vaccines reported breakthrough infections [87]. In the mRNA vaccine and adenovirus vector vaccine studies, efficacy after primary vaccination was reported as breakthrough COVID-19 infections [45,69,97,98], hospitalizations [95,97], and deaths [100]. Efficacy after booster vaccination was reported as breakthrough COVID-19 infections, hospitalizations, deaths [89], and hospitalizations or deaths due to breakthrough infections [90]. In the inactivated virus vaccine and adenovirus vaccine studies, one study reported breakthrough COVID-19 infections after primary vaccination [103]. In the mRNA vaccine, inactivated virus vaccine, and adenovirus vector vaccine studies, efficacy was reported as breakthrough COVID-19 infections and hospitalizations after primary vaccination [88,94]. Breakthrough COVID-19 infections and deaths after booster vaccination were reported in only one study [94].

mRNA vaccination, either primary or booster, has been found to have a protective effect on breakthrough infections where the risk of getting infections after vaccination is lower compared with the unvaccinated group [33]. According to Bieber et al., patients with autoimmune rheumatic disease who received a third booster of mRNA vaccination had lower SARS-CoV-2 infection rates [33]. However, Kim et al. observed both patients and healthy controls to have SARS-CoV-2 omicron breakthrough infections after a third dose of vaccination [95]. Mena-Vázquez et al. also reported that patients who were not infected with SARS-CoV-2 received vaccinations more frequently. Moreover, COVID-19-infected patients took rituximab and glucocorticoids more frequently [63].

Symptomatic breakthrough COVID-19 infections among patients and in a healthy control group were reported in two studies after the participants had had a primary inactivated COVID-19 vaccination [76,77], although only one patient required hospitalization and no patients died [77]. Non-severe infections were reported after a mean period of fourteen weeks from full vaccination, where half of the infected participants were patients with negative total anti-SARS-CoV-2 IgG antibodies and neutralizing antibodies [83].

Studies in which autoimmune patients received an mRNA or adenovirus vector vaccine reported a higher hospitalization rate in the unvaccinated group compared with the vaccinated group, as well as a higher rate of severe COVID-19 cases, which appeared less frequently in third-dose-vaccinated patients than in second-dose-vaccinated patients and an unvaccinated group [89]. Breakthrough infections were also more frequent in patients on strongly impairing immunosuppressants, including anti-CD20 combination therapy, sphingosine 1-phosphate modulators, and mycophenolate mofetil therapy, as opposed to patients on other immunosuppressants [97].

According to the results from a study on inactivated and adenovirus vaccines, the strongest predictor of breakthrough infections is the absence of an antibody response. Vaccine platform and mycophenolate mofetil were found to be the other breakthrough infection predictors [103]. Patients with autoimmune disease receiving Covaxin showed higher rates of breakthrough infection than those receiving the AstraZeneca vaccine [103]. Another result from a study on adenovirus vector vaccines reported that there was no significant difference in the frequency of breakthrough infections between patients who received a second dose of vaccine after 4–6 weeks versus 10–14 weeks [87]. Furthermore, results from a study reporting on autoimmune patients given mRNA, inactivated virus, or adenovirus vector vaccines showed no breakthrough infections in patients vaccinated with mRNA. Meanwhile, inactivated-virus-vaccinated patients had a higher percentage of breakthrough infections after full vaccination than adenovirus-vector-vaccinated patients, although the difference was not significant [88].

3.3.2. Immunogenicity

There were 54 studies reporting immunogenicity: 27 studies on mRNA vaccines (13 studies on Pfizer/BioNTech [30,31,34,36,38,39,43,46,54,59,61,67,68], 1 study on Moderna [101], and 13 studies on Pfizer/BioNTech or Moderna [8,35,41,44,47,50–53,60,62,65,71]); 9 studies on inactivated virus vaccines using CoronaVac [72–77,79,80,83]; 2 studies on adenovirus vector vaccines [84,87]; 12 studies on mRNA and adenovirus vector vaccines (5 studies on Pfizer/BioNTech, Moderna, Vaxzevria, or Janssen [55,57,66,95,98], 2 studies on Pfizer/BioNTech, Moderna, or Vaxzevria [56,58], 3 studies on Pfizer/BioNTech or Vaxzevria [91–93], 1 study on Moderna or Vaxzevria [96], and 1 study on Pfizer/BioNTech (BNT162b2), Moderna (mRNA-1273), or Janssen (Ad26.COV2.S)) [85]; 1 study on inactivated virus vaccines and adenovirus vector vaccines using Covaxin or AstraZeneca [103]; 3 studies on mRNA vaccines, inactivated vaccines, and adenovirus vector vaccines using Pfizer, Coronavac, or Vaxzevria [81], Pfizer, Coronavac, Sinopharm or Vaxzevria [82], and Pfizer, Moderna, Sinopharm, Sputnik, and AstraZeneca [99]. Immunogenicity was determined by measuring antibody titers [8,30,31,34–36,38,39,41,43,46,47,50–54,56–62,65–68,75,76,80–85,87,92,93,96, 98,99,101,103], seroconversion [8,30,31,35,36,39,41,43,44,46,47,52–57,60,61,65,66,71–77,80,82,84, 92,95,97–99,101], neutralization antibodies [38,41,47,50,51,53,54,58,62,68,72–77,80,81,83,84,95, 99,101,103], T-cell response [41,43,46,53–57,60,62,65,66,68,71,81–83,91,95,99,101], lymphocyte count [31,93], IgA titer [38,50,58,85,93], IgG avidity [51,54], B-cell counts [43,49,53,56–58,93], T-cell counts [55,58,62,93], and IgM titer [93].

Patients with autoimmune diseases who received CoronaVac had neutralizing antibodies and neutralizing activity lower than in the control group [72,74,75] as well as lower seroconversion [74,75]. Factors associated with poor immunogenicity were older age, obesity, and use of prednisone, biologics, and immunosuppressants [74,75]. Another study on patients given CoronaVac also found that mycophenolate and prednisone were related to reduced seroconversion, whereas hydroxychloroquine caused seroconversion to rise [77]. In another study on Pfizer, CoronaVac, Sinopharm, and Vaxzevria vaccination, anti-RBD titers were lower in the inactivated vaccine group, followed by Vaxzevria, then Vaxzevria or Pfizer [82]. The inactivated vaccine was also associated with the lowest humoral response, whereas the adenovirus-vectored/mRNA vaccine was associated with the highest humoral response [82].

Patients with multiple sclerosis who received the Pfizer vaccine while being treated with anti-CD20 therapy [54], fingolimod continuation [31], and other immunosuppressants [34] had lower IgG titers compared with untreated patients or patients who discontinued the therapy. In comparison with healthy controls, patients with autoimmune neurological disorder who had received the Pfizer or Moderna vaccines had decreased seroconversion rates [34] and anti-S1 IgG [53,60] and anti-S(RBD) specific IgG levels [52].

In comparison with healthy controls or patients not receiving immunotherapy, patients receiving anti-CD20 [52,53,60], fingolimod [52,53], azathioprine [52], and steroid therapy [52] exhibited lower levels of anti-S1 IgG and anti-S(RBD) specific IgG. Lower seroconversion rates were observed in multiple sclerosis patients receiving anti-CD20 or sphingosine 1-phosphate receptor modulators who were given the Pfizer, Moderna, or AstraZeneca vaccines compared with other disease-modifying therapies or untreated patients [66].

Additional research on the Pfizer or Vaxzevria vaccine indicated that seroconversion and anti-S IgG levels after the second dose were significantly lower in patients with autoimmune disease than in the control group and that this was associated with B-cell depletion at the time of vaccination [92]. Rituximab was significantly associated with no antibody vaccine response after adjusting for diagnosis and hydroxychloroquine, according to research in patients with SLE and RA who received the Pfizer vaccine [39]. A study of patients with SLE who had been given the Pfizer vaccination found that mycophenolate and methotrexate treatment were associated with a drastically diminished BNT162b2 antibody response [68]. Another study on the Moderna and Vaxzevria vaccines showed that individuals given hydroxychloroquine, low-dose steroid, methotrexate, and/or sulfasalazine therapy had significantly lower anti-SARS-CoV-2 spike IgG titers than those who were not on these therapies [96].

Studies of patients with autoimmune or autoinflammatory diseases, including RA, SLE, Sjogren syndrome, Behcet's disease, polymyalgia rheumatica, connective tissue disease, vasculitis, adult-onset Still's disease, and sarcoidosis who received the Pfizer vaccination showed lower seroconversion [46,61], anti S1/S2 IgG [36,38,43,61], neutralization [38,43], total IgA [38,43], and anti-RBD IgG [61] than the control group. The lowest antibody titers were detected in patients with antineutrophil cytoplasmic-antibody-associated vasculitis (AAV) and idiopathic inflammatory myopathy/myositis (IIM), while the highest titers were detected in SLE and RA patients [37]. Another study showed that antibody titers were also reduced with two or more immunosuppressants in combination therapy [61]. Studies in patients with systemic autoimmune disease, RA, SLE, inflammatory bowel disease, Sjogren syndrome, autoimmune hepatitis, psoriatic arthritis, IIM, sarcoidosis, and vasculitis who received the Pfizer and Moderna vaccines found that their anti-S IgG titers were lower than those of the control group, and these differences were particularly significant [8,35,50,51] in those who were receiving B-cell-depleting therapies, prednisone, JAK inhibitors, antimetabolites [47], TNFi [51], mycophenolate, and calcineurin inhibitors [44]. Moreover, compared with the control group, anti-RBD titers were lower in patients [41]. The differences remained significant in individuals receiving treatment with rituximab and belimumab [41]. According to another study, Ab levels and neutralization efficacy against variants of concern in anti-TNF-treated patients were substantially lower than in healthy controls, and by three months following the second dose of the vaccination they were undetectable against Omicron [71].

Seroconversion was considerably higher among Pfizer vaccine recipients when doses were given less than a month apart compared with AstraZeneca recipients, and tendencies towards higher antibody levels in vaccine responders were seen when either vaccine was given using short-interval dosing [93]. A study by Mehta et al. on the AstraZeneca vaccine showed that diabetes mellitus and vaccine interval were significantly associated with anti-RBD antibody titer [87]. A delayed (10–14 weeks) second dose of AstraZeneca vaccine was associated with a higher antibody titer [87]. A study by Ahmed et al. on AstraZeneca and Covaxin revealed that Covaxin and methotrexate treatment were associated with lower antibody titers [103]. Another study that focused on Vaxzevria vaccination in patients with autoimmune inflammatory rheumatic diseases revealed that single-dose-vaccinated patients who had had prior COVID-19 infections showed significantly higher seroconversion and neutralization activity than those who had received a double-dose vaccine [84].

In a study that focused on CoronaVac vaccination, neutralizing antibodies in RA patients on methotrexate therapy were lower than in the control group [73,83], as was the seroconversion rate [73]. Prednisone and mycophenolate usage were both highly linked to a negative NAb [83].

A study on mRNA and inactivated virus vaccines reported that IFN-γ and anti-RBD Abs levels have a slight but significant positive correlation [43]. Another study on mRNA, inactivated virus, and adenovirus vector vaccines reported that neutralizing anti-RBD-specific antibodies and the percentage of positive anti-RBD antibody responses were higher in participants vaccinated with mRNA vaccine compared with inactivated virus and adenovirus vaccines [99]. Additionally, patients who received the adenovirus vector or mRNA vaccines had a higher proportion of TNF-a-producing CD4+ T-cells upon SARS-CoV-2 antigen exposure compared with those who received the inactivated virus vaccine [99].

A third booster dose of mRNA or adenovirus vector vaccine after a primary inactivated vaccine produced a significant humoral and cellular immune response in SLE patients with inactive disease maintaining immunosuppressive treatment [81]. However, another study found that, after booster vaccination, neutralization responses against the Omicron variant were significantly lower in patients than in the healthy control group [92]. Certain medications, such as TNFi, aCD20-BCD- and fingolimod, antimetabolites, and calcineurin inhibitors were able to impair humoral and cellular responses, especially in autoimmune patients [51,53,81]. For instance, Achiron et al. found that a fingolimod continuation group had lower IgG titers than a fingolimod discontinuation group even at 3 months after the third vaccine dose [31]. In addition, anti-BA.2 neutralizing antibodies were not detectable in TNFi-treated patients [51]. Meyer et al. found that patients taking fingolimod failed to develop either humoral or $CD4^+$ T cellular immune responses [53]. In contrast, Meyer et al. also reported that untreated patients showed an increase in anti-S1 IgG, neutralizing capacity, RBD- and S2-specific B cells, and spike-specific T cells after their first booster [53]. Lastly, however, a booster dose, particularly from an mRNA or viral vector vaccine, enhanced strong cellular immune responses, though responses were weaker in patients taking antimetabolites or calcineurin inhibitors [81].

3.3.3. Safety

Following primary vaccination in mRNA vaccine studies, autoimmune relapse was reported as a safety outcome in 12 studies [29,40,60,63,68,70,80–82,86,88,99]; local symptoms in 11 studies [8,29,35–37,40,50,52,60,67,68]; systemic symptoms in 13 studies [8,34–37,40,50, 52,60,63,67,68,101]; and other symptoms in 12 studies [29,35,36,40,50,52,60,64,88,100–102]. Following booster vaccination in mRNA vaccine studies, autoimmune relapse was reported as a safety outcome in three studies [32,34,42] and local and systemic symptoms were also reported in one study [34]. Among the inactivated vaccine studies, following primary vaccination, autoimmune relapse was reported as a safety outcome in two studies [79,80] and local symptoms and systemic symptoms in six studies [72,73,75–78].

Among the mRNA vaccine and adenovirus vector vaccine studies, following primary vaccination, two studies reported autoimmune relapse after vaccination [20,69]. Local symptoms were reported as a safety outcome in three studies [45,69,72]. Systemic symptoms after primary mRNA and adenovirus vector vaccinations were reported in three studies [45,48,102]. Other symptoms were described in one study [82]. Meanwhile, among the inactivated virus vaccine and adenovirus vector vaccine studies, only one study reported autoimmune relapses and local and systemic symptoms as safety outcomes [86]. Among the mRNA vaccine, inactivated virus vaccine, and adenovirus vector vaccine studies, following primary vaccination, autoimmune relapse was reported as a safety outcome in two studies [88,94]; local symptoms in three studies [88,94]; systemic symptoms in four studies [82,88,94,99]; and other adverse events in two studies [88,94].

Patients who had been vaccinated with an mRNA vaccine reported no difference in relapse incidence before and after vaccination [8,52,68]. De Santis et al. and Ferri et al. also reported that, in the majority of cases, vaccine-related adverse effects were mild, and incidence rates were comparable in autoimmune patients and healthy controls with no differences based on current medications [8,101]. Mild cases, such as headache, occurred more frequently in SLE and cryoglobulinemic vasculitis patients, while pain at the injection site did in systemic vasculitis patients [8]. Moyon et al. found no related serious adverse events caused by vaccination [68]. Most of the relapse cases had significantly higher disease activity scores when compared with patients without post-vaccination relapses [40]. Additionally, De Santis et al. did not find any differences between patients with and without serum responses or in the prevalence of vaccine-related side effects [101]. In terms of booster vaccinations, a study reported more than 10% ITP exacerbations among ITP patients after booster vaccinations [32].

Patients who had been vaccinated with an inactivated virus vaccine were reported to have no moderate or severe adverse events [76,77]. Medeiros-Ribeiro et al. reported that overall reactions, such as arthralgia, back pain, malaise, nausea, and sweating, were more frequently and significantly found to occur in patients with autoimmune rheumatic disease than in a control group [75]. In patients with RA, myalgia and vertigo were significantly more frequent in those patients who were stopping methotrexate therapy at the time of receiving their second vaccination [73]. Headaches had a higher prevalence in patients with systemic autoimmune myopathies compared with healthy controls after a first dose of inactivated vaccine [76]. Autoimmune flare was also detected more frequently in a methotrexate-stopping RA patient group in comparison with a methotrexate-maintaining group at day 69 after vaccine administration [73].

Studies on mRNA and adenovirus vector vaccines reported that there was no difference in self-reported side effects between patients with neuroinflammatory diseases and a control group, whether after first vaccine dose or second vaccine dose, even after adjusting for age, BMI, and comorbidities [48]. Epstein et al. also reported that younger age was associated with an increased rate of reported side effects, whereas patients on high-efficacy therapy were associated with a lower risk of reported side effects [48]. The high-efficacy therapies referred to were therapies using ocrelizumab, rituximab, ofatumumab, alemtuzumab, cladribine, fingolimod, ozanimod, siponimod, and natalizumab [48]. Headaches were more common in patients with neuroinflammatory disease after mRNA vaccination than adenovirus vector vaccination, although no significant differences were observed [48]. Additionally, patients on high-efficacy therapy had a significantly lower rate of reported side effects compared with patients not on medication at the time of vaccination [48]. In terms of flare, there were no differences observed regarding age, comorbidities, number of autoimmune diseases associated, and years from disease diagnosis to the year prior to vaccination [69]. There were no significant differences in flare-up development among Cominarty, Vaxzevria, and Spikevax [69].

Additionally, studies on mRNA, inactivated virus, and adenovirus vector vaccines reported significantly more injection site pain in patients receiving AstraZeneca or Pfizer vaccination than in those who received inactivated vaccination, followed by fatigue and fever [82]. Another study on a third booster dose with an mRNA or viral vector vaccine following inactivated virus vaccination in SLE patients revealed more reactogenicity after the booster dose than the initial CoronaVac vaccination, but this was mild and no SLE flare was reported [81].

3.4. Meta-Analysis

For meta-analysis, we included 20 studies that compared the efficacy, immunogenicity, and safety of COVID-19 vaccines between patients with autoimmune diseases and healthy controls. There were six studies for efficacy [72,75–77,97,98], 18 studies for immunogenicity [8,35,36,43,50,52,54,61,62,65,72,74–77,80,83,101], and four studies for safety that could be included [72,75–77]. These studies were on inactivated vaccine, mRNA vaccine, and mRNA/adenovirus vector vaccine. All studies were non-randomized studies and on primary doses (two doses) of COVID-19 vaccine. Meta-analysis could not be done from the RCTs because there were only three RCTs [31,73,96] in our systematic review and only one RCT comparing the efficacy, immunogenicity or safety of the COVID-19 vaccine (primary dose) among patients with autoimmune disease (multiple sclerosis) and healthy controls [31].

3.4.1. Efficacy

Six studies were included to evaluate the efficacy of COVID-19 vaccines in patients with autoimmune diseases. Four studies used the inactivated virus vaccine [72,75–77], whereas the other two studies used mRNA and adenovirus vector vaccines [97,98] Breakthrough COVID-19 infections were used to assess vaccine efficacy.

Based on Figure 2, the overall effect on breakthrough COVID-19 infection after receipt of a COVID-19 inactivated virus vaccine was in favor of the healthy controls. The combined risk ratio was 1.93 (95% CI: 1.14–3.29, I^2 = 0%), and the difference was statistically significant (p = 0.02). According to the GRADE system, the certainty of the evidence on breakthrough COVID-19 infections after inactivated vaccination was moderate (Supplementary Materials, Table S3). Four studies included in this meta-analysis involved patients with various autoimmune diseases: SLE, systemic autoimmune myopathies, and other autoimmune diseases. Patients involved in these four studies received various immunosuppressive treatments: steroids, methotrexate, hydroxychloroquine, mycophenolate mofetil, azathioprine, biologic agents, and others.

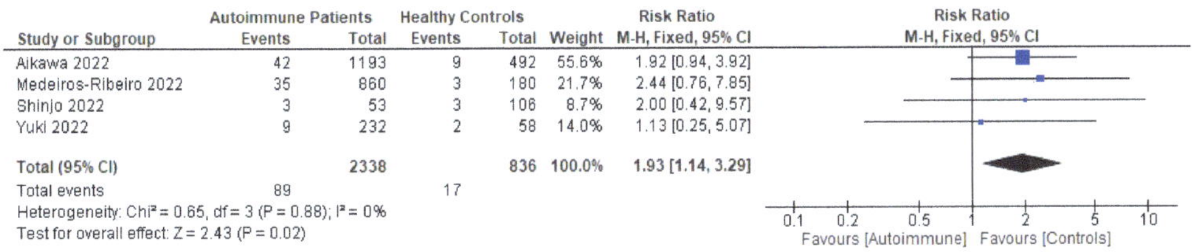

Figure 2. Breakthrough COVID-19 infections after receiving primary doses (two doses) of COVID-19 inactivated vaccine. Blue squares represent effect sizes for a single study, and black rhombus represent pooled results for all studies [72,75–77].

We also analyzed the combined risk ratio for breakthrough infections after mRNA or adenovirus vector vaccination, but no statistically significant difference was observed (RR = 0.97; 95% CI: 0.85–1.11; I^2 = 0%) (Figure 3). According to the GRADE system, the certainty of the evidence on breakthrough COVID-19 infections after mRNA or adenovirus vector vaccination was moderate (Supplementary Materials). Three studies included in this meta-analysis involved patients with various autoimmune diseases: SLE, rheumatoid arthritis, spondiloarthopathy, vasculitis, and others. Patients involved in these four studies received various immunosuppressive treatments: steroids, methotrexate, hydroxychloroquine, leflunomide, mycophenolate mofetil, azathioprine, biologic agents, and others. Subgroup analysis regarding autoimmune diagnosis and treatment could not be done because of limited studies or a lack of subgroup data.

Figure 3. Breakthrough COVID-19 infections from studies using mRNA and adenovirus viral vector COVID 19 vaccines. Blue squares represent effect sizes for a single study, and black rhombus represent pooled results for all studies [97,98].

3.4.2. Immunogenicity

Eighteen studies were included in the meta-analysis to evaluate the immunogenicity of COVID-19 vaccines in patients with autoimmune disease compared with healthy controls [8,35,36,43,50,52,54,61,62,65,72,74–77,80,83,101]. Studies included in this meta-analysis involved patients with various autoimmune diseases: multiple sclerosis, systemic autoimmune diseases, and other autoimmune diseases. Patients involved in these studies received various immunosuppressive treatments: steroids, methotrexate, hydroxychloroquine, mycophenolate mofetil, azathioprine, biologic agents, and others.

Eleven studies were on mRNA vaccines [8,35,36,43,50,52,54,61,62,65,101] and seven studies [72,74–77,80,83] on inactivated vaccines. Seroconversion, proportion of neutralizing antibodies (NAb) positive, log total antibody (TAb) titer, and neutralizing activity were analyzed.

As shown in Figure 4, seven studies reported TAb titers after mRNA vaccination. Patients with autoimmune disease showed significantly lower log TAb (log BAU/mL) titers than healthy controls. Heterogeneity was low (SMD = −0.11, 95% CI = −0.2–0.02, I^2 = 0%). According to the GRADE system, the certainty of the evidence on TAb after mRNA vaccination was high (Supplementary Materials, Table S3).

	Autoimmune Patients			Healthy Controls				Std. Mean Difference
Study or Subgroup	Mean	SD	Total	Mean	SD	Total	Weight	IV, Fixed, 95% CI
De Santis 2022	3.58	3.58	219	3.82	3.69	45	7.4%	-0.07 [-0.39, 0.25]
Ferri 2021	2.71	2.93	478	2.97	2.73	502	48.6%	-0.09 [-0.22, 0.03]
Furer 2021	2.54	2.38	686	2.75	2.33	121	20.4%	-0.09 [-0.28, 0.10]
Furer 2022	2.12	2.31	108	2.75	2.33	122	11.3%	-0.27 [-0.53, -0.01]
Geisen 2021	3.31	3.09	26	3.43	3.04	42	3.2%	-0.04 [-0.53, 0.45]
Malipiero 2021	2.07	2.44	5	3.6	3.27	108	0.9%	-0.47 [-1.37, 0.43]
Mandl 2022	3.06	3.24	82	3.23	2.8	82	8.1%	-0.06 [-0.36, 0.25]
Total (95% CI)			**1604**			**1022**	**100.0%**	**-0.11 [-0.20, -0.02]**
Heterogeneity: Chi² = 2.47, df = 6 (P = 0.87); I² = 0%								
Test for overall effect: Z = 2.43 (P = 0.02)								

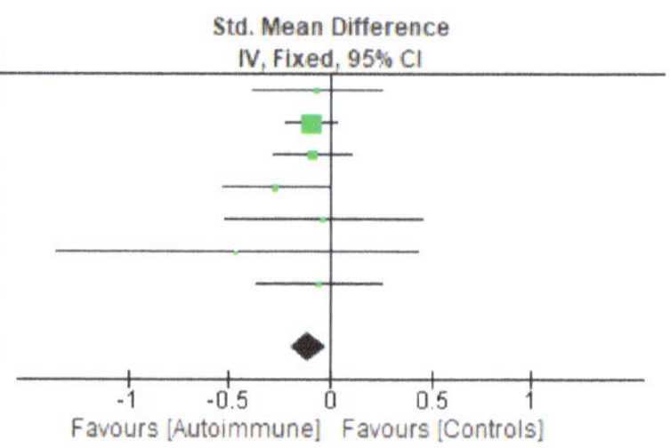

Figure 4. Log TAb titer after mRNA vaccination. Green squares represent effect sizes for a single study, and black rhombus represent pooled results for all studies [8,35,36,43,50,61,101].

As shown in Figure 5, five studies reported Tab titers after inactivated vaccination. Patients with autoimmune disease showed significantly lower log Tab (log BAU/mL) titers compared with healthy controls. Heterogeneity was considerably low (SMD = −0.10, 95% CI = −0.19–0.00, I^2 = 43%). According to the GRADE system, the certainty of the evidence on TAb titer after inactivated vaccination was high (Supplementary Materials, Table S3).

	Autoimmune Patients			Healthy Controls				Std. Mean Difference
Study or Subgroup	Mean	SD	Total	Mean	SD	Total	Weight	IV, Fixed, 95% CI
Aikawa 2022	2.17	2.32	628	2.11	2.04	314	50.5%	0.03 [-0.11, 0.16]
Balcells 2022	1.71	1.46	41	2.08	1.3	65	6.0%	-0.27 [-0.66, 0.12]
Medeiros-Ribeiro 2021	1.85	1.97	859	2.24	2.13	179	35.6%	-0.20 [-0.36, -0.03]
Seree-Aphinan 2021	1.97	1.9	14	2.27	1.91	18	1.9%	-0.15 [-0.85, 0.55]
Shinjo 2022	1.68	1.57	37	2.19	1.38	79	6.0%	-0.35 [-0.74, 0.04]
Total (95% CI)			**1579**			**655**	**100.0%**	**-0.10 [-0.19, 0.00]**
Heterogeneity: Chi² = 7.00, df = 4 (P = 0.14); I² = 43%								
Test for overall effect: Z = 1.96 (P = 0.05)								

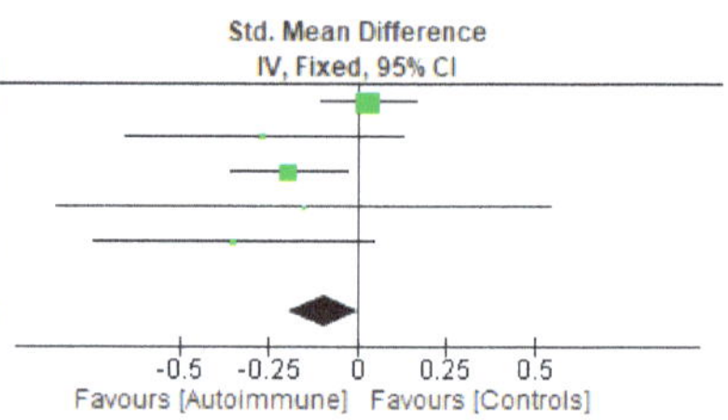

Figure 5. Log TAb titer after inactivated vaccination. Green squares represent effect sizes for a single study, and black rhombus represent pooled results for all studies [72,75,76,80,83].

As shown in Figure 6, 11 studies reported IgG seroconversion after mRNA vaccination compared with healthy controls. IgG Seroconversion after mRNA vaccination was significantly lower among patients with autoimmune disease than healthy controls. Heterogeneity was high (RR = 0.82, 95% CI = 0.75–0.90, I^2 = 97%). According to the GRADE system, the certainty of the evidence on IgG seroconversion after mRNA vaccination was moderate (Supplementary Materials).

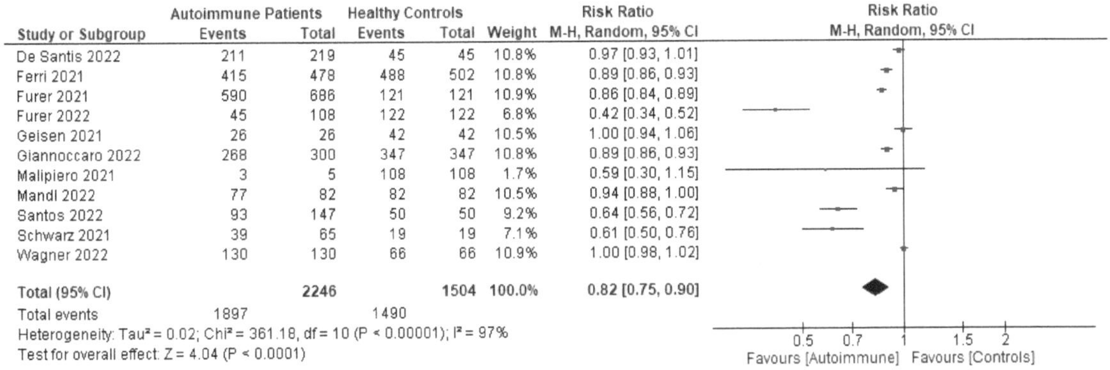

Figure 6. IgG seroconversion after mRNA vaccination. Blue squares represent effect sizes for a single study, and black rhombus represent pooled results for all studies [8,35,36,43,50,52,54,61,62,65,101].

As shown in Figure 7, seven studies reported IgG seroconversion after inactivated vaccination compared with healthy controls. IgG seroconversion after inactivated vaccination was significantly lower among patients with autoimmune disease than healthy controls. Heterogeneity was considerably high (RR = 0.77, 95% CI = 0.71–0.84, I^2 = 86%). According to the GRADE system, the certainty of the evidence on IgG seroconversion after mRNA vaccination was moderate (Supplementary Materials, Table S3).

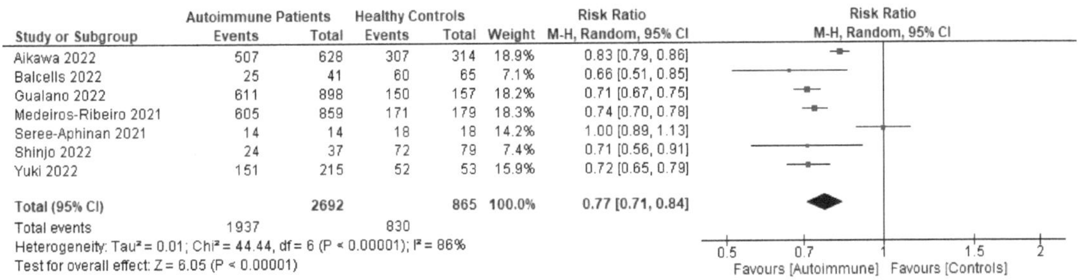

Figure 7. IgG seroconversion after inactivated vaccination. Blue squares represent effect sizes for a single study, and black rhombus represent pooled results for all studies [72,74–77,80,83].

As shown in Figure 8, three studies reported neutralizing antibodies after mRNA vaccination. Patients with autoimmune disease showed a lower proportion of positive NAb than healthy controls, but the difference was not statistically significant. Heterogeneity was high (RR = 0.79, 95% CI = 0.54–1.14, I^2 = 97%). According to the GRADE system, the certainty of the evidence on neutralizing antibodies after mRNA vaccination was very low (Supplementary Materials, Table S3).

As shown in Figure 9, seven studies reported neutralizing antibodies after inactivated vaccination. Patients with autoimmune disease had a significantly lower proportion of positive NAb than healthy controls. Heterogeneity was considerably low (RR = 0.71, 95% CI = 0.68–0.74, I^2 = 37%). According to the GRADE system, the certainty of the evidence on neutralizing antibodies after inactivated vaccination was high (Supplementary Materials, Table S3).

As shown in Figure 10, six studies reported neutralizing activity after inactivated vaccination. Patients with autoimmune disease showed lower mean neutralizing activity after inactivated vaccination than healthy controls, but the result was not statistically significant. Heterogeneity was high (SMD = −0.52, 95% CI = −1.34–0.30, I^2 = 98%). According to the

GRADE system, the certainty of the evidence on neutralizing antibodies after the first dose of vaccine was very low (Supplementary Materials, Table S3).

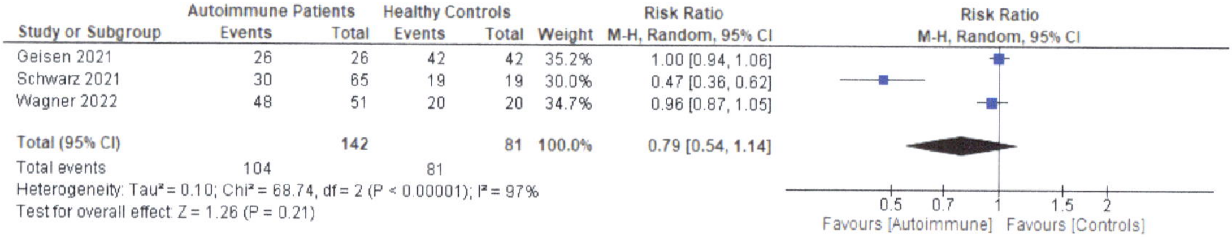

Figure 8. Proportion of neutralizing antibodies positive after mRNA vaccination. Blue squares represent effect sizes for a single study, and black rhombus represent pooled results for all studies [50,54,62].

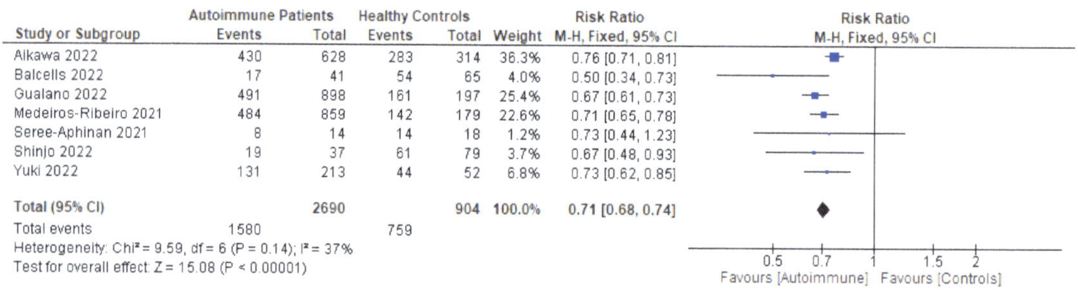

Figure 9. Proportion of neutralizing antibodies positive after inactivated vaccination. Blue squares represent effect sizes for a single study, and black rhombus represent pooled results for all studies [72,74–77,80,83].

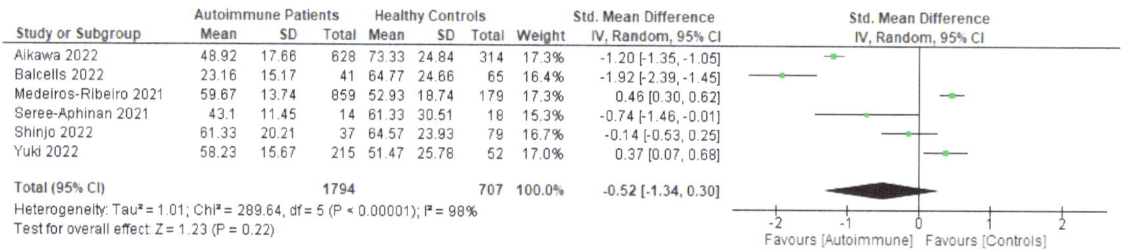

Figure 10. Neutralizing activity after inactivated vaccination. Green squares represent effect sizes for a single study, and black rhombus represent pooled results for all studies [72,75–77,80,83].

3.4.3. Safety

Four studies were eligible for pooling of vaccine-associated adverse events, including local and systemic adverse events. All included studies were on inactivated COVID-19 vaccines [72,75–77]. Four studies included in this meta-analysis involved patients with various autoimmune diseases: SLE, systemic autoimmune myopathies, and other autoimmune diseases. Patients involved in these four studies received various immunosuppressive treatments: steroids, methotrexate, hydroxychloroquine, mycophenolate mofetil, azathioprine, biologic agents, and others.

We observed that the combined risk ratio for local adverse events after a first dose of COVID-19 inactivated vaccine was 1.26 (95% CI: 1.05–1.51; $I^2 = 0\%$) (Figure 11). Patients with autoimmune diseases had a statistically significant ($p = 0.01$) risk of local adverse events after receiving a first dose of COVID-19 inactivated vaccine in comparison with healthy controls. According to the GRADE system, the certainty of the evidence on local adverse events after first dose COVID-19 inactivated vaccine was high (Supplementary Materials, Table S3).

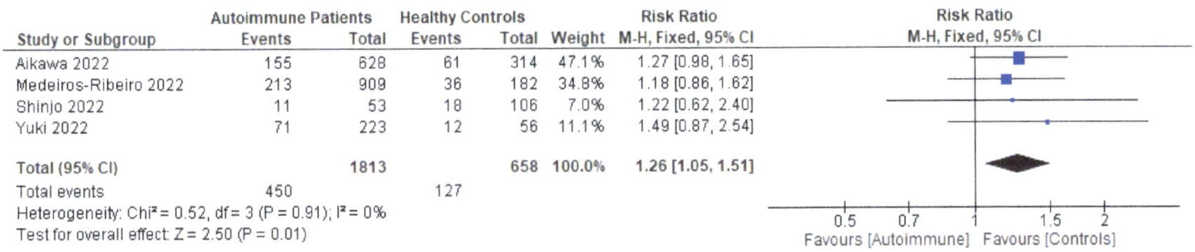

Figure 11. Local adverse events after receiving a first dose of COVID-19 inactivated vaccine. Blue squares represent effect sizes for a single study, and black rhombus represent pooled results for all studies [72,75–77].

We observed that the combined risk ratio for local adverse events after a second dose COVID-19 inactivated vaccine was 1.11 (95% CI: 0.91–1.35; $I^2 = 1\%$) (Figure 12). Patients with autoimmune diseases had a higher risk of local adverse events than healthy controls after receiving a second dose of COVID-19 inactivated vaccine, but the difference was not statistically significant ($p = 0.31$). According to the GRADE system, the certainty of the evidence for local adverse events after a second dose of COVID-19 inactivated vaccine was high (Supplementary Materials, Table S3).

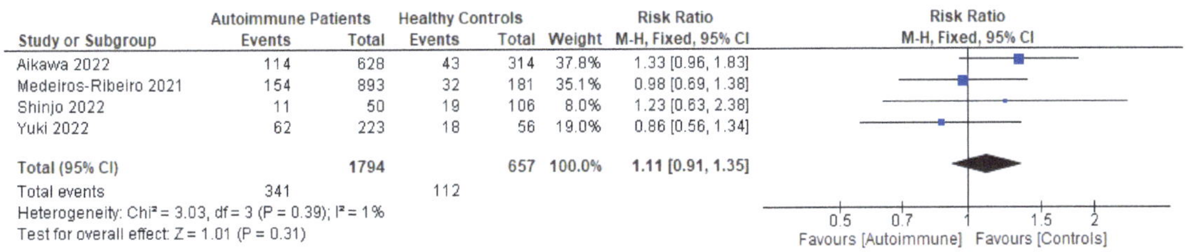

Figure 12. Local adverse events after receiving a second dose of COVID-19 inactivated vaccine. Blue squares represent effect sizes for a single study, and black rhombus represent pooled results for all studies [72,75–77].

We observed that the combined risk ratio for systemic adverse events after a first dose of COVID-19 inactivated vaccine was 1.31 (95% CI: 1.15–1.48; $I^2 = 0\%$). Patients with autoimmune diseases had a statistically significant ($p < 0.0001$) risk of systemic adverse events after receiving a first dose of COVID-19 inactivated vaccine in comparison with healthy controls (Figure 13). According to the GRADE system, the certainty of the evidence on systemic adverse events after a first dose of COVID-19 inactivated vaccine was high (Supplementary Materials, Table S3).

The combined risk ratio for systemic adverse events after a second dose of COVID-19 inactivated vaccine was 1.13 (Figure 14), but no statistically significant difference was observed (95% CI: 0.88–1.45; $I^2 = 62\%$). According to the GRADE system, the certainty of the evidence on local adverse events was moderate (Supplementary Materials, Table S3).

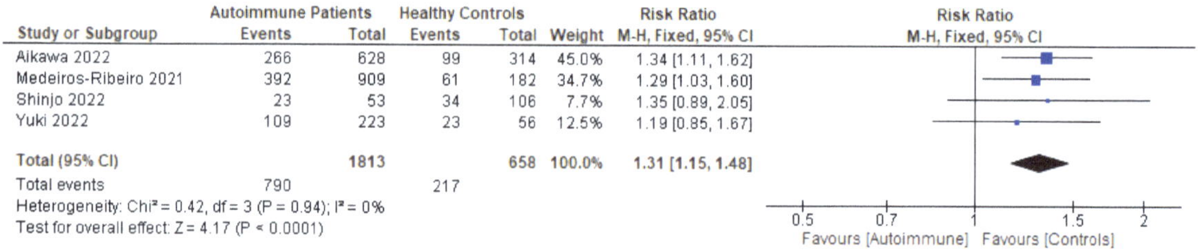

Figure 13. Systemic adverse events after receiving a first dose of COVID-19 inactivated vaccine. Blue squares represent effect sizes for a single study, and black rhombus represent pooled results for all studies [72,75–77].

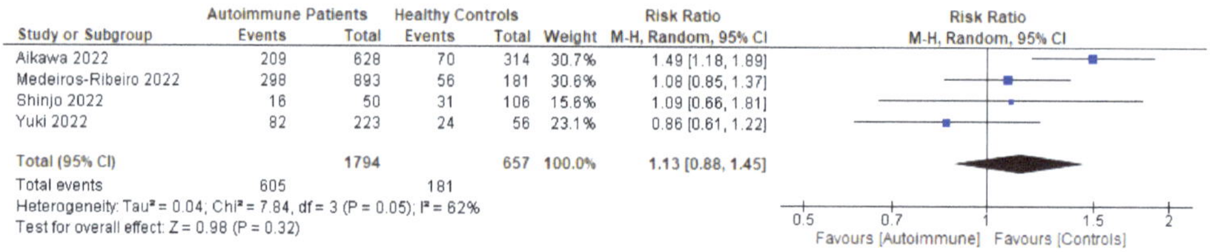

Figure 14. Systemic adverse events after receiving a second dose of COVID-19 inactivated vaccine. Blue squares represent effect sizes for a single study, and black rhombus represent pooled results for all studies [72,75–77].

3.5. Publication Bias

We used a funnel plot to assess publication bias for a meta-analysis involving more than 10 studies: IgG seroconversion after mRNA vaccination (Supplementary Materials, Figure S2). The funnel plot was asymmetrical, which could indicate that there was publication bias.

4. Discussion

There are some issues regarding COVID-19 vaccination in autoimmune patients, such as how autoimmune medications might affect the efficacy and immunogenicity of the vaccines and possible adverse reactions following COVID-19 vaccination. Therefore, the efficacy, immunogenicity, and safety of COVID-19 vaccines in autoimmune patients were the primary outcomes in this systematic review and meta-analysis.

Only a few studies were identified that addressed all three outcomes. In the meta-analysis, we compared efficacy, immunogenicity, and safety between patients with autoimmune diseases and healthy controls. Because of the heterogeneity of the studies, we only had non-randomized studies that could be used for this purpose. We also could not conduct a meta-analysis on booster (third-dose) COVID-19 vaccination due to limited studies sharing similar outcomes and interventions.

Regarding the efficacy of COVID-19 vaccination, our meta-analysis showed that the risk of breakthrough COVID-19 infection significantly increased in patients with autoimmune diseases compared with healthy controls after receipt of an inactivated virus vaccine. On the other hand, a meta-analysis with studies using mRNA or adenovirus vectors did not show significant differences in breakthrough infections among patients with autoimmune disease compared with healthy controls. Breakthrough COVID-19 infection can be related to viral profile, host factors (comorbidities, immunosuppressive drugs), and vaccine platform or dose. The mRNA vaccine platform shows stronger neutralizing antibody and T cell responses compared with other vaccine platforms [104].

Ahmed et al. reported that only small numbers of breakthrough infections occurred in patients with autoimmune diseases after they received either an inactivated or adenovirus vector vaccine [103]. Furer et al. observed no symptomatic COVID-19 infections in patients with autoimmune diseases, and only one subject in the healthy control group was diagnosed with a breakthrough COVID-19 infection after a second dose of mRNA vaccine during the study follow-up [35]. Moreover, Stalman et al. reported breakthrough COVID-19 infections after mRNA or adenovirus vector vaccine in both autoimmune patients and healthy controls, with no differences in the trends in the incidence rates [97]. Kim et al. also reported breakthrough infections after booster vaccination with an mRNA vaccine in subjects given an mRNA or adenovirus vector vaccine as their primary COVID-19 vaccination, but the result was not significantly different between patients with autoimmune disease and healthy controls (healthcare workers) [95].

Studies included in a meta-analysis of breakthrough infections after mRNA or adenovirus vector vaccination involved patients with various diagnoses and treatments for autoimmune diseases. Patel et al. and Paik et al. explained that increased breakthrough infections were associated with the use of multiple immunomodulatory therapies, such as methotrexate, mycophenolate mofetil, anti-CD20, and TNF inhibitors [105,106]. A study by Bieber et al. also showed higher doses of steroids and higher proportions of patients given TNF alpha inhibitors, rituximab, and calcineurin inhibitors among cases of breakthrough COVID-19 infection [33].

Regarding the immunogenicity of the vaccine, our meta-analyses showed that patients with autoimmune diseases had reduced total antibody (TAb) titers, IgG seroconversion, and neutralizing antibodies after COVID-19 inactivated vaccination compared with healthy controls. Patients with autoimmune diseases also showed reduced TAb titers and IgG seroconversion after COVID-19 mRNA vaccination compared with healthy controls. A study by Kim et al. on mRNA vaccine boosters showed that limited neutralization of the Omicron variant in the sera of patients with autoimmune disease could contribute to a shorter median time between third-dose vaccination and the time of breakthrough infection compared with a control group [95].

Patients with autoimmune diseases showed noticeably different humoral responses following vaccination, which may be attributed to the use of B-cell-depleting agents, antimetabolites, glucocorticoids, other immunosuppressive drugs, and waning immunity [106]. This was proven by Ferri et al. in a study that showed an increased prevalence of non-responders to vaccines in patients with systemic autoimmune disease treated with glucocorticoids, mycophenolate mofetil, and rituximab [8]. So et al. found that impaired humoral response in SLE patients significantly correlated with the use of mycophenolate and the type of vaccine, especially inactivated virus vaccines in comparison with mRNA vaccines [107]. Paik et al. reported that B-cell-depleting agents, antimetabolites, glucocorticoids, and combination immunosuppressive therapy achieved significantly lower seroconversion, while immunomodulators, such as hydroxychloroquine and intravenous globulin, did not reduce antibody titers [106]. However, patients treated with hydroxychloroquine, combined with other therapies such as methotrexate and/or sulfasalazine, still had significantly lower anti-SARS-CoV spike IgG antibody titers than those who did not receive such a combination [96].

In terms of the safety of vaccination, the overall estimate from the meta-analysis showed a significantly higher risk for patients with autoimmune disease experiencing local and systemic adverse events after a first dose of COVID-19 inactivated vaccine in comparison with healthy controls; however, no statistically significant difference after a second dose of vaccine was observed. Higher frequencies of adverse events were reported among seropositive patients than in seronegative patients and healthy controls [72]. No moderate or severe adverse events related to the vaccine were reported [72,75–77]. Vaccine-related adverse events after the inactivated COVID-19 vaccine, especially systemic symptoms, were fewer than those reported with the mRNA vaccine [75].

In our systematic review, flare (worsening of autoimmune disease activity) was observed in more than 10% of patients with SLE after primary mRNA vaccination [70], and in patients with hematologic autoimmune diseases including immune-mediated thrombotic thrombocytopenic purpura and immune thrombocytopenia after a booster mRNA vaccination [32,42]. Meanwhile, this occurred in less than 5% of patients with multiple sclerosis [29] and with systemic autoimmune diseases including cryoglobulinemic vasculitis, rheumatoid arthritis, systemic lupus erythematosus, and systemic sclerosis after primary mRNA vaccination [8]. For the other vaccine types, flare was observed in 7% of autoimmune skin disease patients after primary inactivated virus vaccine and in less than 5% of SLE and autoimmune rheumatic disease patients [77,80]. Other adverse events, such as face tingling, herpes reactivation, bleeding, and urinary tract infection, also occurred in a small number of patients, together with severe adverse events such as high blood pressure, immune thrombocytopenic purpura, myocarditis, and death [29,80,101]. However, causal and temporal relationships between vaccine administration and adverse events or worsening disease activity following vaccination were difficult to determine due to limited data and the lack of a specific analysis of the causal relationship.

Based on our qualitative findings, breakthrough infections occurred less frequently in autoimmune patients after a booster dose. Autoimmune patients still had lower humoral and cellular responses even after having a third vaccine dose. Most of the patients were on immunosuppressant therapy, while untreated patients had better humoral and cellular responses. These findings support some previous evidence regarding the effects of booster vaccination. Regardless of the lower antibody titers in autoimmune patients, a potential increase in titer could be achieved after administering a third dose of vaccine, though the titer was still lower compared with a healthy control group. Evidence from a study by Joudeh et al. indicates that a booster vaccine dose is associated with a higher seroconversion rate, particularly in patients with a history of COVID-19 infection [108]. Further evidence comes from Cardelli, et al., who showed that a time-dependent decrease in protective antibody titer was restored after receipt of a booster dose. After a booster dose, five of nine non-responders developed adequate anti-RBD and neutralizing antibody titers. Three of them reduced their dose of or discontinued mycophenolate mofetil or azathioprine therapy before booster administration [109]. In addition, in terms of efficacy and safety, Dreyer at al. found no relapse activity or breakthrough infections after the third dose of vaccine [34].

This study has several limitations. First, the number of studies used to combine the efficacy, immunogenicity, and safety findings was relatively small. Second, considering that only one RCT was available comparing patients with autoimmune diseases and healthy controls after a primary dose of COVID-19 vaccine, we only included non-randomized studies. Third, since we only included a small number of studies in our meta-analysis, we might have significant publication bias. However, we also included pre-printed studies in our systematic review to reduce the possibility of this bias. Fourth, the variety of autoimmune diagnoses and immunosuppressive treatments could have an impact on the outcome of COVID-19 vaccination. This could affect our meta-analysis, and we could not address this by subgroup analysis due to the limited studies available.

5. Conclusions

In conclusion, from this meta-analysis, we found that patients with autoimmune diseases showed significantly more breakthrough COVID-19 infections and lower total antibody (TAb) titers, IgG seroconversion, and neutralizing antibodies after inactivated COVID-19 vaccination compared with healthy controls. They also had more local and systemic adverse events after a first dose of inactivated vaccination compared with healthy controls, but this result was not seen after a second dose. Patients with autoimmune diseases also showed significantly lower TAb titers and IgG seroconversion after COVID-19 mRNA vaccination compared with healthy controls.

A second dose of vaccine was, however, found to be important, since it is associated with improved antibody titers and seroconversion. It is important to consult a healthcare provider before taking a vaccine, since immunosuppressants might affect the immunogenicity of vaccines. Additionally, the administration of third doses of COVID-19 vaccines should be considered due to improved seroprotection in these patients.

Supplementary Materials: The following supporting information can be downloaded at: https://www.mdpi.com/article/10.3390/vaccines11091456/s1, Table S1: Search strategy; Table S2: Summarize of research articles; Table S3: GRADE asessement; Figure S1: (A): Quality assessment of RCT studies, (B): Quality assessment of non-RCT studies, (C): Quality assessment of case-series studies, (D): Quality assessment of cross-sectional studies; Figure S2: Funnel plot of IgG seroconversion after mRNA vaccination.

Author Contributions: Conceptualization, A.W., A.S.H., S.M., Y.O., C.Y., S.K., E.Y. and S.D.; methodology, R.R., Y.O., C.Y. and K.H.; software, R.R. and C.Y.; validation, A.S.H., K.H, C.Y., S.K. and E.Y.; formal analysis, A.W., A.S.H., R.R., S.M., M.I.H., Y.O., C.Y. and K.H.; investigation, A.W., A.S.H., R.R., S.M., M.I.H., Y.O., C.Y. and A.N.L.S.; resources, Y.O., C.Y., S.K., E.Y. and S.D.; data curation, A.W., R.R. and M.I.H.; writing—original draft preparation, A.W., A.S.H., R.R., S.M. and M.I.H.; writing—review and editing, Y.O., C.Y., K.H, S.K., E.Y. and S.D.; visualization, R.R.; supervision, S.K., E.Y. and S.D.; project administration, R.R.; funding acquisition, A.W., A.S.H., Y.O. and C.Y. All authors have read and agreed to the published version of the manuscript.

Funding: This systematic review received study funding from the University of Indonesia, grant number NKB-745/UN2.RST/HKP.05.00/2022.

Institutional Review Board Statement: Not applicable.

Informed Consent Statement: Not applicable.

Data Availability Statement: The datasets generated and analyzed during the current study are available from the corresponding author upon reasonable request.

Acknowledgments: We express our appreciation to Annisa Layalia Widjanarko, who helped with data collection.

Conflicts of Interest: The authors declare no conflict of interest.

References

1. WHO. WHO Coronavirus (COVID-19) Dashboard | WHO Coronavirus (COVID-19) Dashboard with Vaccination Data. 2022. Available online: https://covid19.who.int/ (accessed on 3 May 2022).
2. Centers for Disease Control Prevention (CDC) People with Certain Medical, Conditions | CDC. Available online: https://www.cdc.gov/coronavirus/2019-ncov/need-extra-precautions/people-with-medical-conditions.html (accessed on 2 July 2022).
3. Velikova, T.; Georgiev, T. SARS-CoV-2 vaccines and autoimmune diseases amidst the COVID-19 crisis. *Rheumatol. Int.* **2021**, *41*, 509–518. [CrossRef]
4. Soy, M.; Keser, G.; Atagunduz, P.; Mutlu, M.Y.; Gunduz, A.; Koybaşi, G.; Bes, C. A practical approach for vaccinations including COVID-19 in autoimmune/autoinflammatory rheumatic diseases: A non-systematic review. *Clin. Rheumatol.* **2021**, *40*, 3533–3545. [CrossRef] [PubMed]
5. Kaur, S.P.; Gupta, V. COVID-19 Vaccine: A comprehensive status report. *Virus Res.* **2020**, *288*, 198114. [CrossRef]
6. World Health Organization (WHO). COVID-19 Vaccines. 2021. Available online: https://www.who.int/emergencies/diseases/novel-coronavirus-2019/covid-19-vaccines (accessed on 2 July 2021).
7. Furer, V.; Rondaan, C.; Heijstek, M.W.; Agmon-Levin, N.; Assen, S.; Van Bijl, M.; Breedveld, F.C.; D'Amelio, R.; Dougados, M.; Kapetanovic, M.C.; et al. 2019 update of EULAR recommendations for vaccination in adult patients with autoimmune inflammatory rheumatic diseases. *Ann. Rheum. Dis.* **2020**, *79*, 39–52. [CrossRef] [PubMed]
8. Ferri, C.; Ursini, F.; Gragnani, L.; Raimondo, V.; Giuggioli, D.; Foti, R.; Caminiti, M.; Olivo, D.; Cuomo, G.; Visentini, N.; et al. Impaired immunogenicity to COVID-19 vaccines in autoimmune systemic diseases. High prevalence of non-response in different patients' subgroups. *J. Autoimmun.* **2021**, *125*, 102744. [CrossRef] [PubMed]
9. Widhani, A.; Sukmana, N. Vaksinasi pada penyakit autoimun. In *Dalam*; Djauzi, S., Rengganis, I., Sundoro, J., Koesnoe, S., Soegiarto, G., Maria, S., Eds.; Pedoman Imunisasi Pada Orang Dewasa; Satgas Imunisasi Dewasa PAPDI: Jakarta, Indonesia, 2017.
10. Toussirot, É.; Bereau, M. Vaccination and induction of autoimmune diseases. *Inflamm. Allergy Drug Targets* **2015**, *14*, 94–98. [CrossRef]
11. Soriano, A.; Nesher, G.; Shoenfeld, Y. Predicting post-vaccination autoimmunity: Who might be at risk? *Pharmacol. Res.* **2015**, *92*, 18–22. [CrossRef]
12. Shoenfeld, Y. Corona (COVID-19) time musings: Our involvement in COVID-19 pathogenesis, diagnosis, treatment and vaccine planning. *Autoimmun. Rev.* **2020**, *19*, 102538. [PubMed]
13. Vojdani, A.; Kharrazian, D. Potential antigenic cross-reactivity between SARS-CoV-2 and human tissue with a possible link to an increase in autoimmune diseases. *Clin. Immunol.* **2020**, *217*, 108480. [CrossRef]

14. Voysey, M.; Clemens, S.A.C.; Madhi, S.A.; Weckx, L.Y.; Folegatti, P.M.; Aley, P.K.; Angus, B.; Baillie, V.L.; Barnabas, S.L.; Bharat, Q.E.; et al. Safety and efficacy of the ChAdOx1 nCoV-19 vaccine (AZD1222) against SARS-CoV-2: An interim analysis of four randomised controlled trials in Brazil, South Africa, and the UK. *Lancet* **2021**, *397*, 99–111. [CrossRef]
15. Pottegård, A.; Lund, L.C.; Karlstad, Ø.; Dahl, J.; Andersen, M.; Hallas, J.; Lidegaard, Ø.; Tapi, G.; Gulseth, H.L.; Ruiz, P.L.; et al. Arterial events, venous thromboembolism, thrombocytopenia, and bleeding after vaccination with Oxford-AstraZeneca ChAdOx1-S in Denmark and Norway: Population based cohort study. *BMJ* **2021**, *373*, n1114. [CrossRef]
16. Schultz, N.H.; Sørvoll, I.H.; Michelsen, A.E.; Munthe, L.A.; Lund-Johansen, F.; Ahlen, M.T.; Wiedmann, M.; Aamodt, A.; Skattør, T.H.; Tjønnfjord, G.E.; et al. Thrombosis and Thrombocytopenia after ChAdOx1 nCoV-19 Vaccination. *N. Engl. J. Med.* **2021**, *384*, 2124–21230. [CrossRef]
17. Greinacher, A.; Thiele, T.; Warkentin, T.E.; Weisser, K.; Kyrle, P.A.; Eichinger, S. Thrombotic Thrombocytopenia after ChAdOx1 nCov-19 Vaccination. *N. Eng. J. Med.* **2021**, *384*, 2092–2101. [CrossRef]
18. Scully, M.; Singh, D.; Lown, R.; Poles, A.; Solomon, T.; Levi, M.; Goldblatt, D.; Kotoucek, P.; Thomas, W.; Lester, W. Pathologic Antibodies to Platelet Factor 4 after ChAdOx1 nCoV-19 Vaccination. *N. Eng. J. Med.* **2021**, *384*, 2202–2211. [CrossRef] [PubMed]
19. Bril, F.; Diffalha SAl Dean, M.; Fettig, D.M. Autoimmune hepatitis developing after coronavirus disease 2019 (COVID-19) vaccine: Causality or casualty? *J. Hepatol.* **2021**, *75*, 222–224. [CrossRef] [PubMed]
20. Lee, E.J.; Cines, D.B.; Gernsheimer, T.; Kessler, C.; Michel, M.; Tarantino, M.D.; Semple, J.W.; Arnold, D.M.; Godeau, B.; Lambert, M.P.; et al. Thrombocytopenia following Pfizer and Moderna SARS-CoV-2 vaccination. *Am. J. Hematol.* **2021**, *96*, 534–537. [CrossRef]
21. Hazlewood, G.S.; Pardo, J.P.; Barnabe, C.; Schieir, O.; Barber, C.E.H.; Bernatsky, S.; Colmegna, I.; Hitcon, C.; Loeb, M.; Mertz, D.; et al. Canadian rheumatology association recommendation for the use of COVID-19 vaccination for patients with autoimmune rheumatic diseases. *J. Rheumatol.* **2021**, *48*, 1330–1339. [CrossRef] [PubMed]
22. Moher, D.; Liberati, A.; Tetzlaff, J.; Altman, D.G. Preferred Reporting Items for Systematic Reviews and Meta-Analyses: The PRISMA Statement. *Ann. Intern. Med.* **2009**, *151*, 264–269. [CrossRef]
23. Infantino, M.; Pieri, M.; Nuccetelli, M.; Grossi, V.; Lari, B.; Tomassetti, F.; Calugi, G.; Pancani, S.; Benucci, M.; Casprini, P.; et al. The WHO International Standard for COVID-19 serological tests: Towards harmonization of anti-spike assays. *Int. Immunopharmacol.* **2021**, *100*, 108095. [CrossRef] [PubMed]
24. Valcourt, E.J.; Manguiat, K.; Robinson, A.; Chen, J.C.; Dimitrova, K.; Philipson, C.; Lamoreux, L.; McLachlan, E.; Schiffman, Z.; Drebot, M.A.; et al. Evaluation of a commercially-available surrogate virus neutralization test for severe acute respiratory syndrome coronavirus-2 (SARS-CoV-2). *Diagn. Microbiol. Infect. Dis.* **2021**, *99*, 115294. [CrossRef] [PubMed]
25. Wan, X.; Wang, W.; Liu, J.; Tong, T. Estimating the sample mean and standard deviation from the sample size, median, range and/or interquartile range. *BMC Med. Res. Methodol.* **2014**, *14*, 135. [CrossRef]
26. Higgins, J.; Thomas, J.; Chandler, J.; Cumpston, M.; Li, T.; Page, M.; Welch, V.A. Cochrane Handbook for Systematic Reviews of Interventions, Version 6.3; Updated February 2022; Cochrane: 2022. Available online: https://training.cochrane.org/handbook (accessed on 23 February 2023).
27. Wells, G.A.; Shea, B.; O'Connel, D.; Peterson, J.; Welch, V.; Losos, M.; Tugwell, P. The Newcastle-Ottawa Scale (NOS) for Assessing the Quality of Nonrandomised Studies in Meta-Analyses. Available online: https://www.ohri.ca/programs/clinical_epidemiology/oxford.asp (accessed on 1 April 2023).
28. The National Institutes of Health. Study Quality Assessment Tools. 2013. Available online: https://www.nhlbi.nih.gov/health-topics/study-quality-assessment-tools (accessed on 1 April 2023).
29. Achiron, A.; Dolev, M.; Menascu, S.; Zohar, D.N.; Dreyer-Alster, S.; Miron, S.; Shirbint, E.; Megalashvili, D.; Fletcher, S.; Givon, U.; et al. COVID-19 vaccination in patients with multiple sclerosis: What we have learnt by February 2021. *Mult. Scler.* **2021**, *27*, 864–870. [CrossRef]
30. Achiron, A.; Mandel, M.; Dreyer-Alster, S.; Harari, G.; Magalashvili, D.; Sonis, P.; Dolev, M.; Menascu, S.; Fletcher, S.; Falb, R.; et al. Humoral immune response to COVID-19 mRNA vaccine in patients with multiple sclerosis treated with high-efficacy disease-modifying therapies. *Ther. Adv. Neurol. Disord.* **2021**, *14*, 17562864211022581. [CrossRef] [PubMed]
31. Achiron, A.; Mandel, M.; Gurevich, M.; Dreyer-Alster, S.; Magalashvili, D.; Sonis, P.; Dolev, M.; Menascu, S.; Harari, G.; Fletcher, S.; et al. Immune response to the third COVID-19 vaccine dose is related to lymphocyte count in multiple sclerosis patients treated with fingolimod. *J. Neurol.* **2021**, *269*, 2286–2292. [CrossRef] [PubMed]
32. Aharoni, M.; Leader, A.; Shochat, T.; Raanani, P.; Spectre, G. Exacerbation of immune thrombocytopenia following initial and booster vaccination with Pfizer-BioNTech COVID-19 vaccine. *Platelets* **2022**, *33*, 781–786. [CrossRef] [PubMed]
33. Bieber, A.; Sagy, I.; Novack, L.; Brikman, S.; Abuhasira, R.; Ayalon, S.; Novofastovski, I.; Abu-Shakra, M.; Mader, R. BNT162b2 mRNA COVID-19 vaccine and booster in patients with autoimmune rheumatic diseases: A national cohort study. *Ann. Rheum. Dis.* **2022**, *81*, 1028–1035. [CrossRef] [PubMed]
34. Dreyer-Alster, S.; Menascu, S.; Mandel, M.; Shirbint, E.; Magalashvili, D.; Dolev, M.; Fletcher, S.; Givon, U.; Guber, D.; Stern, Y.; et al. COVID-19 vaccination in patients with multiple sclerosis: Safety and humoral efficacy of the third booster dose. *J. Neurol. Sci.* **2022**, *434*, 120155. [CrossRef]

35. Furer, V.; Eviatar, T.; Zisman, D.; Peleg, H.; Paran, D.; Levartovsky, D.; Zisapel, M.; Elalouf, O.; Kaufman, I.; Meidan, R.; et al. Immunogenicity and safety of the BNT162b2 mRNA COVID-19 vaccine in adult patients with autoimmune inflammatory rheumatic diseases and in the general population: A multicentre study. *Ann. Rheum. Dis.* **2021**, *80*, 1330–1338. [CrossRef] [PubMed]
36. Furer, V.; Eviatar, T.; Zisman, D.; Peleg, H.; Braun-Moscovici, Y.; Balbir-Gurman, A.; Paran, D.; Levartovsky, D.; Zisapel, M.; Elalouf, O.; et al. Predictors of Immunogenic Response to the BNT162b2 mRNA COVID-19 Vaccination in Patients with Autoimmune Inflammatory Rheumatic Diseases Treated with Rituximab. *Vaccines* **2022**, *10*, 901. [CrossRef]
37. Furer, V.; Zisman, D.; Kibari, A.; Rimar, D.; Paran, Y.; Elkayam, O. Herpes zoster following BNT162b2 mRNA COVID-19 vaccination in patients with autoimmune inflammatory rheumatic diseases: A case series. *Rheumatology* **2021**, *60*, SI90–SI95. [CrossRef] [PubMed]
38. Lustig, Y.; Sapir, E.; Regev-Yochay, G.; Cohen, C.; Fluss, R.; Olmer, L.; Indebaum, V.; Mandelboim, M.; Doolman, R.; Amit, S.; et al. BNT162b2 COVID-19 vaccine and correlates of humoral immune responses and dynamics: A prospective, single-centre, longitudinal cohort study in health-care workers. *Lancet Respir. Med.* **2021**, *9*, 999–1009. [CrossRef] [PubMed]
39. Ammitzbøll, C.; Bartels, L.E.; Andersen, J.B.; Vils, S.R.; Mistegård, C.E.; Johannsen, A.D.; Hermansen, M.F.; Thomsen, M.K.; Erikstrup, C.; Hauge, E.; et al. Impaired Antibody Response to the BNT162b2 Messenger RNA Coronavirus Disease 2019 Vaccine in Patients With Systemic Lupus Erythematosus and Rheumatoid Arthritis. *ACR Open Rheumatol.* **2021**, *3*, 622–628. [CrossRef] [PubMed]
40. Dinoto, A.; Gastaldi, M.; Iorio, R.; Marini, S.; Damato, V.; Farina, A.; Zoccarato, M.; Sechi, E.; Pinna, F.; Maniscalco, G.T.; et al. Safety profile of SARS-CoV-2 vaccination in patients with antibody-mediated CNS disorders. *Mult. Scler. Relat. Disord.* **2022**, *63*, 103827. [CrossRef] [PubMed]
41. Fabris, M.; de Marchi, G.; Domenis, R.; Caponnetto, F.; Guella, S.; Secco, C.D.; Cabas, N.; De Vita, S.; Beltrami, A.P.; Curcio, F.; et al. High T-cell response rate after COVID-19 vaccination in belimumab and rituximab recipients. *J. Autoimmun.* **2022**, *129*, 102827. [CrossRef] [PubMed]
42. Giuffrida, G.; Markovic, U.; Condorelli, A.; Calagna, M.; Grasso, S.; Duminuco, A.; Riccobene, C.; Pelle, A.C.; Zanghi, G.; Di Raimondo, F. Relapse of immune-mediated thrombotic thrombocytopenic purpura following mRNA COVID-19 vaccination: A prospective cohort study. *Haematologica* **2022**, *107*, 2661–2666. [CrossRef]
43. Malipiero, G.; Moratto, A.; Infantino, M.; D'Agaro, P.; Piscianz, E.; Manfredi, M.; Grossi, V.; Benvenuti, E.; Bulgaresi, M.; Benucci, M.; et al. Assessment of humoral and cellular immunity induced by the BNT162b2 SARS-CoV-2 vaccine in healthcare workers, elderly people, and immunosuppressed patients with autoimmune disease. *Immunol. Res.* **2021**, *69*, 576–583. [CrossRef]
44. Zecca, E.; Rizzi, M.; Tonello, S.; Matino, E.; Costanzo, M.; Rizzi, E.; Casciaro, G.F.; Manfredi, G.F.; Acquaviva, A.; Gagliardi, I.; et al. Ongoing Mycophenolate Treatment Impairs Anti-SARS-CoV-2 Vaccination Response in Patients Affected by Chronic Inflammatory Autoimmune Diseases or Liver Transplantation Recipients: Results of the RIVALSA Prospective Cohort. *Viruses* **2022**, *14*, 1766. [CrossRef]
45. Gerosa, M.; Schioppo, T.; Argolini, L.M.; Sciascia, S.; Ramirez, G.A.; Moroni, G.; Sinico, R.A.; Bonelli, G.; Alberici, F.; Mescia, F.; et al. The Impact of Anti-SARS-CoV-2 Vaccine in Patients with Systemic Lupus Erythematosus: A Multicentre Cohort Study. *Vaccines* **2022**, *10*, 663. [CrossRef]
46. Costa, C.; Scozzari, G.; Migliore, E.; Galassi, C.; Ciccone, G.; Ricciardelli, G.; Scarmozzino, A.; Angeloe, L.; Cassoni, P.; Cavallo, R. Cellular Immune Response to BNT162b2 mRNA COVID-19 Vaccine in a Large Cohort of Healthcare Workers in a Tertiary Care University Hospital. *Vaccines* **2022**, *10*, 1031. [CrossRef]
47. Deepak, P.; Kim, W.; Paley, M.A.; Yang, M.; Carvidi, A.B.; El-Qunni, A.A. Glucocorticoids and B Cell Depleting Agents Substantially Impair Immunogenicity of mRNA Vaccines to SARS-CoV-2. *medRXiv* **2021**. [CrossRef]
48. Epstein, S.; Xia, Z.; Lee, A.J.; Dahl, M.; Edwards, K.; Levit, E.; Longbrake, E.E.; Perrone, C.; Kavak, K.; Weinstock-Guttman, B.; et al. Vaccination Against SARS-CoV-2 in Neuroinflammatory Disease: Early Safety/Tolerability Data. *Mult. Scler. Relat. Disord.* **2022**, *57*, 103433. [CrossRef] [PubMed]
49. Sakano, T.; Bittner, E.A.; Chang, M.G. Severe COVID pneumonia and undetectable B cells after vaccination in patients previously treated with rituximab: A case series. *Postgrad. Med.* **2022**, *134*, 239–243. [CrossRef]
50. Geisen, U.M.; Berner, D.K.; Tran, F.; Sümbül, M.; Vullriede, L.; Ciripoi, M.; Reid, H.M.; Schaffarzyk, A.; Longardt, A.C.; Franzenburg, J.; et al. Immunogenicity and safety of anti-SARS-CoV-2 mRNA vaccines in patients with chronic inflammatory conditions and immunosuppressive therapy in a monocentric cohort. *Ann. Rheum. Dis.* **2021**, *80*, 1306–1311. [CrossRef]
51. Geisen, U.M.; Rose, R.; Neumann, F.; Ciripoi, M.; Vullriede, L.; Reid, H.M.; Berner, D.K.; Bertoglio, F.; Hoff, P.; Hust, M.; et al. The long term vaccine-induced anti-SARS-CoV-2 immune response is impaired in quantity and quality under TNFα blockade. *J. Med. Virol.* **2022**, *94*, 5780–5789. [CrossRef] [PubMed]
52. Giannoccaro, M.P.; Vacchiano, V.; Leone, M.; Camilli, F.; Zenesini, C.; Panzera, I.; Balboni, A.; Tappata, M.; Borghi, A.; Salvi, F.; et al. Difference in safety and humoral response to mRNA SARS-CoV-2 vaccines in patients with autoimmune neurological disorders: The ANCOVAX study. *J. Neurol.* **2022**, *269*, 4000–4012. [CrossRef] [PubMed]
53. Meyer-Arndt, L.; Braun, J.; Fauchere, F.; Vanshylla, K.; Loyal, L.; Henze, L.; Kruse, B.; Dingeldey, M.; Jürchott, K.; Mangold, M.; et al. SARS-CoV-2 mRNA vaccinations fail to elicit humoral and cellular immune responses in patients with multiple sclerosis receiving fingolimod. *J. Neurol. Neurosurg. Psychiatry* **2022**, *93*, 960–971. [CrossRef]

54. Schwarz, T.; Otto, C.; Jones, T.C.; Pache, F.; Schindler, P.; Niederschweiberer, M.; Schmidt, F.A.; Drosten, C.; Corman, V.M.; Ruprecht, K. Preserved T cell responses to SARS-CoV-2 in anti-CD20 treated multiple sclerosis. *medRxiv* **2021**, *15*, 17562864221141505. [CrossRef] [PubMed]
55. Arnold, F.; Huzly, D.; Tanriver, Y.; Welte, T. Response to SARS-CoV-2 vaccines in patients receiving B-cell modulating antibodies for renal autoimmune disease. *BMC Infect. Dis.* **2021**, *22*, 734. [CrossRef]
56. Duengelhoef, P.; Hartl, J.; Rüther, D.; Steinmann, S.; Brehm, T.T.; Weltzsch, J.P.; Glaser, F.; Schaub, G.M.; Sterneck, M.; Sebode, M.; et al. SARS-CoV-2 vaccination response in patients with autoimmune hepatitis and autoimmune cholestatic liver disease. *United Eur. Gastroenterol. J.* **2022**, *10*, 319–329. [CrossRef]
57. Krasselt, M.; Wagner, U.; Nguyen, P.; Pietsch, C.; Boldt, A.; Baerwald, C.; Pierer, M.; Seifert, O. Humoral and cellular response to COVID-19 vaccination in patients with autoimmune inflammatory rheumatic diseases under real-life conditions. *Rheumatology* **2022**, *61*, 180–188. [CrossRef]
58. Stefanski, A.L.; Rincon-Arevalo, H.; Schrezenmeier, E.; Karberg, K.; Szelinski, F.; Ritter, J.; Jahrsdorfer, B.; Schrenzenmeier, H.; Ludwig, C.; Sattler, A.; et al. B Cell Numbers Predict Humoral and Cellular Response Upon SARS–CoV-2 Vaccination Among Patients Treated With Rituximab. *Arthritis Rheumatol.* **2022**, *74*, 934–947. [CrossRef] [PubMed]
59. Hills, T.; Paterson, A.; Woodward, R.; Middleton, F.; Carlton, L.H.; McGregor, R.; Barfoot, S.; Ramiah, C.; Whitcombe, A.L.; Zimbron, V.M.; et al. The effect of needle length and skin to deltoid muscle distance in adults receiving an mRNA COVID-19 vaccine. *Vaccine* **2022**, *40*, 4827–4834. [CrossRef]
60. Kornek, B.; Leutmezer, F.; Rommer, P.S.; Koblischke, M.; Schneider, L.; Haslacher, H.; Thalhammer, R.; Zimprich, F.; Zulanher, G.; Bsteh, G.; et al. B Cell Depletion and SARS-CoV-2 Vaccine Responses in Neuroimmunologic Patients. *Ann. Neurol.* **2022**, *91*, 342–352. [CrossRef]
61. Mandl, P.; Tobudic, S.; Haslacher, H.; Karonitsch, T.; Mrak, D.; Nothnagl, T.; Perkmann, T.; Radner, H.; Sautner, J.; Simader, E.; et al. Response to SARS-CoV-2 vaccination in systemic autoimmune rheumatic disease depends on immunosuppressive regimen: A matched, prospective cohort study. *Ann. Rheum. Dis.* **2022**, *81*, 1017–1022. [CrossRef]
62. Wagner, A.; Garner-Spitzer, E.; Schötta, A.M.; Orola, M.; Wessely, A.; Zwazl, I.; Ohradanova-Repic, A.; Weseslindtner, L.; Tatji, G.; Gebetsberger, L.; et al. SARS-CoV-2-mRNA Booster Vaccination Reverses Non-Responsiveness and Early Antibody Waning in Immunocompromised Patients—A Phase Four Study Comparing Immune Responses in Patients with Solid Cancers, Multiple Myeloma and Inflammatory Bowel Disease. *Front. Immunol.* **2022**, *13*, 889138. [CrossRef] [PubMed]
63. Mena-Vázquez, N.; García-Studer, A.; Rojas-Gimenez, M.; Romero-Barco, C.M.; Manrique-Arija, S.; Mucientes, A.; Velloso-Feijoo, M.L.; Godoy-Navarette, F.J.; Morales-Garrido, P.; Redondo-Rodriguez, R.; et al. Importance of Vaccination against SARS-CoV-2 in Patients with Interstitial Lung Disease Associated with Systemic Autoimmune Disease. *J. Clin. Med.* **2022**, *11*, 2437. [CrossRef]
64. Quintanilla-Bordás, C.; Gascón-Gimenez, F.; Alcalá, C.; Payá, M.; Mallada, J.; Silla, R.; Carratalá-Bosca, S.; Gasque-Rubio, R.; Castillo, J.; Casanova, B. Case Report: Exacerbation of Relapses Following mRNA COVID-19 Vaccination in Multiple Sclerosis: A Case Series. *Front. Neurol.* **2022**, *13*, 897275. [CrossRef] [PubMed]
65. Santos, C.S.; Antolin, S.C.; Morales, C.M.; Herrero, J.G.; Alvarez, E.D.; Ortega, F.R.; de Morales, J.G.R. Immune responses to mRNA vaccines against SARS-CoV-2 in patients with immune-mediated inflammatory rheumatic diseases. *RMD Open.* **2022**, *8*, e001898. [CrossRef]
66. Zabalza, A.; Arrambide, G.; Otero-Romero, S.; Pappolla, A.; Tagliani, P.; López-Maza, S.; Cárdenas-Robledo, S.; Esperalba, J.; Fernández-Naval, C.; Martínez-Gallo, M.; et al. Is humoral and cellular response to SARS-CoV-2 vaccine modified by DMT in patients with multiple sclerosis and other autoimmune diseases? *Mult. Scler.* **2022**, *28*, 1138–1145. [CrossRef]
67. Mitsunaga, T.; Ohtaki, Y.; Seki, Y.; Yoshioka, M.; Mori, H.; Suzuka, M.; Mashiko, S.; Takeda, S.; Mashiko, K. Evaluation of the antibody response and adverse reactions of the BNT162b2 vaccine of participants with prior COVID-19 infection in Japan. *medRxiv* **2021**. [CrossRef]
68. Moyon, Q.; Sterlin, D.; Miyara, M.; Anna, F.; Mathian, A.; Lhote, R.; Ghillani-Dalbin, P.; Breillat, P.; Mudumba, S.; de Alba, S.; et al. BNT162b2 vaccine-induced humoral and cellular responses against SARS-CoV-2 variants in Systemic Lupus Erythematosus. *Ann. Rheum. Dis.* **2021**, *81*, 575–583. [CrossRef] [PubMed]
69. Pinte, L.; Negoi, F.; Ionescu, G.D.; Caraiola, S.; Balaban, D.V.; Badea, C.; Mazilu, D.; Dumitrescu, B.; Mateescu, B.; Lonescu, R.; et al. COVID-19 Vaccine Does Not Increase the Risk of Disease Flare-Ups among Patients with Autoimmune and Immune-Mediated Diseases. *J. Pers. Med.* **2021**, *11*, 1283. [CrossRef]
70. Zavala-Flores, E.; Salcedo-Matienzo, J.; Quiroz-Alva, A.; Berrocal-Kasay, A. Side effects and flares risk after SARS-CoV-2 vaccination in patients with systemic lupus erythematosus. *Clin. Rheumatol.* **2022**, *41*, 1349–1357. [CrossRef] [PubMed]
71. Dayam, R.M.; Law, J.C.; Goetgebuer, R.L.; Chao, G.Y.C.; Abe, K.T.; Sutton, M.; Finkelstein, N.; Stempak, J.M.; Pereira, D.; Croitoru, D.; et al. Accelerated waning of immunity to SARS-CoV-2 mRNA vaccines in patients with immune-mediated inflammatory diseases. *JCI Insight* **2022**, *7*, e159721. [CrossRef]
72. Aikawa, N.E.; Kupa, L.V.K.; Pasoto, S.G.; Medeiros-Ribeiro, A.C.; Yuki, E.F.N.; Saad, C.G.S.; Pedrosa, T.; Fuller, R.; Shinjo, S.K.; Sampaio-Barros, P.; et al. Immunogenicity and safety of two doses of the CoronaVac SARS-CoV-2 vaccine in SARS-CoV-2 seropositive and seronegative patients with autoimmune rheumatic diseases in Brazil: A subgroup analysis of a phase 4 prospective study. *Lancet Rheumatol.* **2022**, *4*, 113–124. [CrossRef]

73. Araujo, C.S.R.; Medeiros-Ribeiro, A.C.; Saad, C.G.S.; Bonfiglioli, K.R.; Domiciano, D.S.; Shimabuco, A.Y.; Silvo, M.S.R.; Yuki, E.F.N.; Pasoto, S.G.; Pedrosa, T.; et al. Two-week methotrexate discontinuation in patients with rheumatoid arthritis vaccinated with inactivated SARS-CoV-2 vaccine: A randomised clinical trial. *Ann. Rheum. Dis.* **2022**, *81*, 889–897. [CrossRef]
74. Gualano, B.; Lemes, I.R.; Silva, R.P.; Pinto, A.J.; Mazzolani, B.C.; Smaira, F.I.; Sieczkowska, S.M.; Aikawa, N.E.; Pasoto, S.G.; Medeiros-Ribeiro, A.C.; et al. Association between physical activity and immunogenicity of an inactivated virus vaccine against SARS-CoV-2 in patients with autoimmune rheumatic diseases. *Brain Behav. Immun.* **2022**, *101*, 49–56. [CrossRef] [PubMed]
75. Medeiros-Ribeiro, A.C.; Aikawa, N.E.; Saad, C.G.S.; Yuki, E.F.N.; Pedrosa, T.; Fusco, S.R.G.; Rojo, P.T.; Pereira, R.M.R.; Shinjo, S.K.; Andrade, D.C.O.; et al. Immunogenicity and safety of the CoronaVac inactivated vaccine in patients with autoimmune rheumatic diseases: A phase 4 trial. *Nat. Med.* **2021**, *27*, 1744–1751. [CrossRef]
76. Shinjo, S.K.; de Souza, F.H.C.; Borges, I.B.P.; Dos Santos, A.M.; Miossi, R.; Misse, R.G.; Medeiros-Ribeiro, A.C.; Saad, C.G.S.; Yuki, E.F.N.; Pasoto, S.G.; et al. Systemic autoimmune myopathies: A prospective phase 4 controlled trial of an inactivated virus vaccine against SARS-CoV-2. *Rheumatology* **2022**, *61*, 3351–3561. [CrossRef]
77. Yuki, E.F.N.; Borba, E.F.; Pasoto, S.G.; Seguro, L.P.; Lopes, M.; Saad, C.G.S.; Medeiros-Ribeiro, A.C.; Silva, C.A.; de Andrade, D.C.O.; Kupa, L.D.K.; et al. Impact of Distinct Therapies on Antibody Response to SARS-CoV-2 Vaccine in Systemic Lupus Erythematosus. *Arthritis Care Res.* **2022**, *74*, 562–571. [CrossRef]
78. Chen, J.; Cai, W.; Liu, T.; Zhou, Y.; Jin, Y.; Yang, Y.; Chen, S.; Tang, K.; Li, C. The COVID-19 vaccine: Attitudes and vaccination in patients with autoimmune inflammatory rheumatic diseases. *Rheumatol. Autoimmun.* **2022**, *2*, 82–91. [CrossRef]
79. Huang, L.; Jiang, Z.; Zhou, J.; Chen, Y.; Huang, H. The Effect of Inactivated SARS-CoV-2 Vaccines on TRAB in Graves' Disease. *Front. Endocrinol.* **2022**, *13*, 835880. [CrossRef]
80. Seree-Aphinan, C.; Chanprapaph, K.; Rattanakaemakorn, P.; Setthaudom, C.; Suangtamai, T.; Pomsoong, C.; Ratanapokasatit, Y.; Suchonwanit, P. Inactivated COVID-19 Vaccine Induces a Low Humoral Immune Response in a Subset of Dermatological Patients Receiving Immunosuppressants. *Front. Med.* **2021**, *8*, 769845. [CrossRef]
81. Assawasaksakul, T.; Sathitratanacheewin, S.; Vichaiwattana, P.; Wanlapakorn, N.; Poovorawan, Y.; Kittanamongkolchai, W. Immunogenicity, safety and reactogenicity of a heterogeneous booster following the CoronaVac inactivated SARS-CoV-2 vaccine in patients with SLE: A case series. *RMD Open* **2021**, *7*, e002019. [CrossRef] [PubMed]
82. Assawasaksakul, T.; Lertussavavivat, T.; Sathitratanacheewin, S.; Oudomying, N.; Vichaiwattana, P.; Wanlapakorn, N.; Poorvorawan, Y.; Avihingsanon, Y.; Assawasaksakul, N.; Buranapraditkun, S.; et al. Comparison of Immunogenicity and Safety of Inactivated, Adenovirus-Vectored, and Heterologous Adenovirus-Vectored/mRNA Vaccines in Patients with Systemic Lupus Erythematosus and Rheumatoid Arthritis: A Prospective Cohort Study. *Vaccines* **2022**, *10*, 853. [CrossRef]
83. Balcells, M.E.; Le Corre, N.; Durán, J.; Ceballos, M.E.; Vizcaya, C.; Mondaca, S.; Dib, M.; Rabagliati, R.; Sarmiento, M.; Burgos, P.I.; et al. Reduced Immune Response to Inactivated Severe Acute Respiratory Syndrome Coronavirus 2 Vaccine in a Cohort of Immunocompromised Patients in Chile. *Clin. Infect. Dis.* **2022**, *75*, e594–e602. [CrossRef] [PubMed]
84. Shenoy, P.; Ahmed, S.; Paul, A.; Cherian, S.; Umesh, R.; Shenoy, V.; Vijayan, A.; Babu, S.; Nivin, S.; Thambi, A. Hybrid immunity versus vaccine-induced immunity against SARS CoV2 in Patients with Autoimmune Rheumatic Diseases. *medRxiv* **2021**, *4*, e80–e82. [CrossRef]
85. Timmermann, L.; Globke, B.; Lurje, G.; Schmelzle, M.; Schöning, W.; Öllinger, R.; Pratschke, J.; Bettina, E.; Drosten, C.; Hofmann, J.; et al. Humoral Immune Response following SARS-CoV-2 Vaccination in Liver Transplant Recipients. *Vaccines* **2021**, *9*, 1422. [CrossRef]
86. Cherian, S.; Paul, A.; Ahmed, S.; Alias, B.; Manoj, M.; Santhosh, A.K.; Varghese, D.R.; Krishnan, N.; Shenoy, P. Safety of the ChAdOx1 nCoV-19 and the BBV152 vaccines in 724 patients with rheumatic diseases: A post-vaccination cross-sectional survey. *Rheumatol. Int.* **2021**, *41*, 1441–1445. [CrossRef]
87. Mehta, P.; Paul, A.; Ahmed, S.; Cherian, S.; Panthak, A.; Benny, J.; Shenoy, P. Effectiveness of delayed second dose of AZD1222 vaccine in patients with autoimmune rheumatic disease. *Clin. Rheumatol.* **2022**, *41*, 3537–3542. [CrossRef] [PubMed]
88. Mohanasundaram, K.; Santhanam, S.; Natarajan, R.; Murugesan, H.; Nambi, T.; Chilikuri, B.; Nallasivan, S. Covid-19 vaccination in autoimmune rheumatic diseases: A multi-center survey from southern India. *Int. J. Rheum. Dis.* **2022**, *25*, 1046–1052. [CrossRef] [PubMed]
89. Bakasis, A.D.; Mavragani, C.P.; Voulgari, P.V.; Gerolymatou, N.; Argyropoulou, O.D.; Vlachoyiannopoulos, P.G.; Skopuli, F.N.; Tzioufas, A.G.; Moutsopoulos, H.M. COVID-19: Clinical features and outcomes in unvaccinated 2-dose and 3-dose vaccinated against SARS-CoV-2 patients with systemic autoimmune and autoinflammatory rheumatic diseases. *J. Autoimmun.* **2022**, *131*, 102846. [CrossRef]
90. Efe, C.; Taşçılar, K.; Gerussi, A.; Bolis, F.; Lammert, C.; Ebik, B.; Stättermayer, A.F.; Cengiz, M.; Gökçe, D.T.; Cristoferi, L.; et al. SARS-CoV-2 vaccination and risk of severe COVID-19 outcomes in patients with autoimmune hepatitis. *J. Autoimmun.* **2022**, *132*, 102906. [CrossRef]
91. Kennedy, N.A.; Lin, S.; Goodhand, J.R.; Chanchlani, N.; Hamilton, B.; Bewshea, C.; Nice, R.; Chee, D.; Cummings, J.F.; Fraser, A.; et al. Infliximab is associated with attenuated immunogenicity to BNT162b2 and ChAdOx1 nCoV-19 SARS-CoV-2 vaccines in patients with IBD. *Gut* **2021**, *70*, 1884–1893. [CrossRef]
92. Prendecki, M.; Clarke, C.; Edwards, H.; McIntyre, S.; Mortimer, P.; Gleeson, S.; Martin, P.; Thomson, T.; Randell, P.; Shah, A.; et al. Humoral and T-cell responses to SARS-CoV-2 vaccination in patients receiving immunosuppression. *Ann. Rheum. Dis.* **2021**, *80*, 1322–1329. [CrossRef]

93. Shields, A.M.; Venkatachalam, S.; Shafeek, S.; Paneesha, S.; Ford, M.; Sheeran, T.; Kelly, M.; Qureshi, I.; Salhan, B.; Karim, F.; et al. SARS-CoV-2 vaccine responses following CD20-depletion treatment in patients with haematological and rheumatological disease: A West Midlands Research Consortium study. *Clin. Exp. Immunol.* **2022**, *207*, 3–10. [CrossRef] [PubMed]
94. Machado, P.M.; Lawson-Tovey, S.; Strangfeld, A.; Mateus, E.F.; Hyrich, K.L.; Gossec, L.; Carmona, L.; Rodrigues, A.; Raffeiner, B.; Duarte, C.; et al. Safety of vaccination against SARS-CoV-2 in people with rheumatic and musculoskeletal diseases: Results from the EULAR Coronavirus Vaccine (COVAX) physician-reported registry. *Ann. Rheum. Dis.* **2022**, *81*, 695–709. [CrossRef]
95. Kim, W.J.; Choi, S.H.; Park, J.Y.; Song, J.S.; Chung, J.W.; Choi, S.T. SARS-CoV-2 Omicron escapes mRNA vaccine booster-induced antibody neutralisation in patients with autoimmune rheumatic diseases: An observational cohort study. *Ann. Rheum. Dis.* **2022**, *81*, 1585–1593. [CrossRef] [PubMed]
96. Lin, K.Y.; Hsieh, M.J.; Chang, S.Y.; Ieong, S.M.; Cheng, C.Y.; Sheng, W.H.; Chang, S. Serological response after COVID-19 mRNA-1273 booster dose in immunocompromised patients, Taiwan, July to August 2021. *J. Formos. Med. Assoc.* **2022**, *121*, 2438–2445. [CrossRef] [PubMed]
97. Stalman, E.W.; Wieske, L.; van Dam, K.P.J.; Kummer, L.Y.; Kempen, Z.L.E.; Killestein, J.; Volkers, A.G.; Tas, S.W.; Boekel, L.; Wolbink, G.J.; et al. Breakthrough infections with the SARS-CoV-2 omicron (B.1.1.529) variant in patients with immune-mediated inflammatory diseases. *Ann. Rheum. Dis.* **2022**, *81*, 1757–1766. [CrossRef]
98. Boekel, L.; Steenhuis, M.; Hooijberg, F.; Besten, Y.R.; van Kempen, Z.L.E.; Kummer, L.Y.; van Dam, K.P.J.; Stalman, E.W.; Vogelzang, E.H.; Cristianawati, O.; et al. Antibody development after COVID-19 vaccination in patients with autoimmune diseases in the Netherlands: A substudy of data from two prospective cohort studies. *Lancet Rheumatol.* **2021**, *3*, 778–788. [CrossRef]
99. Szebeni, G.J.; Gémes, N.; Honfi, D.; Szabó, E.; Neuperger, P.; Balog, J.; Nagy, L.I.; Szekanecz, Z.; Puskás, L.G.; Toldi, G.; et al. Humoral and Cellular Immunogenicity and Safety of Five Different SARS-CoV-2 Vaccines in Patients with Autoimmune Rheumatic and Musculoskeletal Diseases in Remission or With Low Disease Activity and in Healthy Controls: A Single Center Study. *Front. Immunol.* **2022**, *13*, 846248. [CrossRef]
100. Lee, E.J.; Beltrami-Moreira, M.; Al-Samkari, H.; Cuker, A.; Diraimo, J.; Gernsheimer, T.; Kruse, A.; Kessler, C.; Kruse, C.; Leavitt, A.D.; et al. SARS-CoV-2 vaccination and ITP in patients with de novo or preexisting ITP. *Blood* **2022**, *139*, 1564–1574. [CrossRef] [PubMed]
101. De Santis, M.; Motta, F.; Isailovic, N.; Clementi, M.; Criscuolo, E.; Clementi, N.; Tonutti, A.; Rodolfi, S.; Barone, E.; Colapietro, F.; et al. Dose-Dependent Impairment of the Immune Response to the Moderna-1273 mRNA Vaccine by Mycophenolate Mofetil in Patients with Rheumatic and Autoimmune Liver Diseases. *Vaccines* **2022**, *10*, 801. [CrossRef]
102. Lonati, P.A.; Bodio, C.; Scavone, M.; Martini, G.; Pesce, E.; Bandera, A.; Lombardi, A.; Gerosa, M.; Franceschini, F.; Tincani, A.; et al. Production of anti-PF4 antibodies in antiphospholipid antibody-positive patients is not affected by COVID-19 vaccination. *RMD Open* **2022**, *8*, 1902. [CrossRef]
103. Ahmed, S.; Mehta, P.; Paul, A.; Anu, S.; Cherian, S.; Shenoy, V.; Nalianda, N.K.; Joseph, S.; Poulose, A.; Shenoy, P. Postvaccination antibody titres predict protection against COVID-19 in patients with autoimmune diseases: Survival analysis in a prospective cohort. *Ann. Rheum. Dis.* **2022**, *81*, 868–874. [CrossRef]
104. Amanatidou, E.; Gkiouliava, A.; Fella, E.; Serafidi, M.; Tsilingiris, D.; Vallianou, N.G.; Karampela, I.; Dalamaga, M. Breakthrough infections after COVID-19 vaccination: Insights, perspectives and challenges. *Metabol. Open* **2022**, *14*, 100180. [CrossRef]
105. Patel, N.J.; Wang, X.; Fu, X.; Kawano, Y.; Cook, C.; Vanni, K.M.M.; Qian, G.; Banasiak, E.; Kowalski, E.; Zhang, Y.; et al. Factors associated with COVID-19 breakthrough infection among vaccinated patients with rheumatic diseases: A cohort study. *Semin. Arthritis Rheum.* **2023**, *58*, 152108. [CrossRef]
106. Paik, J.J.; Sparks, J.A.; Kim, A.H.J. Immunogenicity, breakthrough infection, and underlying disease flare after SARS-CoV-2 vaccination among individuals with systemic autoimmune rheumatic diseases. *Curr. Opin. Pharmacol.* **2022**, *65*, 102243. [CrossRef] [PubMed]
107. So, H.; Li, T.; Chan, V.; Tam, L.S.; Chan, P.K.S. Immunogenicity and safety of inactivated and mRNA COVID-19 vaccines in patients with systemic lupus erythematosus. *Ther. Adv. Musculoskelet. Dis.* **2022**, *14*, 1759720X221089586. [CrossRef] [PubMed]
108. Joudeh, A.I.; Lutf, A.Q.; Mahdi, S.; Tran, G. Efficacy and safety of mRNA and AstraZeneca COVID-19 vaccines in patients with autoimmune rheumatic diseases: A systematic review. *Vaccine* **2023**, *41*, 3812. [CrossRef] [PubMed]
109. Cardelli, C.; Caruso, T.; Tani, C.; Pratesi, F.; Talarico, R.; Di Cianni, F.; Italiano, N.; Laurino, E.; Moretti, M.; Cascarano, G.; et al. COVID-19 mRNA vaccine booster in patients with autoimmune rheumatic diseases. *Rheumatology* **2022**, *61*, 328–330. [CrossRef] [PubMed]

vaccines

MDPI

Article

Results on SARS-CoV-2 mRNA Vaccine Booster from an Open-Label Multicenter Study in Ofatumumab-Treated Participants with Relapsing Multiple Sclerosis

Tjalf Ziemssen [1,*], Eugen Schlegel [2], Marie Groth [3], Benjamin Ettle [3] and Tobias Bopp [4]

1 Department of Neurology, Center of Clinical Neuroscience, Carl Gustav Carus University Clinic, University Hospital of Dresden, Technische Universität Dresden, 01062 Dresden, Germany
2 Zentrum für Neurologische Studien, 57076 Siegen, Germany; schlegel@zns-siegen.de
3 Novartis Pharma GmbH, 90429 Nuremberg, Germany; marie.groth@novartis.com (M.G.); benjamin.ettle@novartis.com (B.E.)
4 Institute for Immunology, University Medical Center, Johannes Gutenberg University, 55131 Mainz, Germany; boppt@uni-mainz.de
* Correspondence: tjalf.ziemssen@uniklinikum-dresden.de

Abstract: Background: Few data exist on how ofatumumab treatment impacts SARS-CoV-2 booster vaccination response. Methods: KYRIOS is an ongoing prospective open-label multicenter study on the response to initial and booster SARS-CoV-2 mRNA vaccination before or during ofatumumab treatment in relapsing MS patients. The results on the initial vaccination cohort have been published previously. Here, we describe 23 patients who received their initial vaccination outside of the study but booster vaccination during the study. Additionally, we report the booster results of two patients in the initial vaccination cohort. The primary endpoint was SARS-CoV-2-specific T-cell response at month 1. Furthermore, serum total and neutralizing antibodies were measured. Results: The primary endpoint was reached by 87.5% of patients with booster before (booster cohort 1, N = 8) and 46.7% of patients with booster during ofatumumab treatment (booster cohort 2, N = 15). Seroconversion rates for neutralizing antibodies increased from 87.5% at baseline to 100.0% at month 1 in booster cohort 1 and from 71.4% to 93.3% in booster cohort 2. Of note, 3 of 4 initially seronegative patients in booster cohort 2 and one seronegative patient in the initial vaccination cohort seroconverted after the booster during ofatumumab treatment. Conclusions: Booster vaccinations increase neutralizing antibody titers in ofatumumab-treated patients. A booster is recommended in ofatumumab-treated patients.

Keywords: COVID-19 vaccination; relapsing multiple sclerosis; ofatumumab; neutralizing antibodies; T-cell responses

Citation: Ziemssen, T.; Schlegel, E.; Groth, M.; Ettle, B.; Bopp, T. Results on SARS-CoV-2 mRNA Vaccine Booster from an Open-Label Multicenter Study in Ofatumumab-Treated Participants with Relapsing Multiple Sclerosis. *Vaccines* **2023**, *11*, 978. https://doi.org/10.3390/vaccines11050978

Academic Editor: Kay Choong See

Received: 16 April 2023
Revised: 4 May 2023
Accepted: 11 May 2023
Published: 13 May 2023

1. Introduction

Health authorities highly recommend vaccination against the severe acute respiratory syndrome coronavirus 2 (SARS-CoV-2) to prevent severe courses of coronavirus disease 2019 (COVID-19). The Robert Koch Institute currently recommends two initial vaccine applications followed by a third vaccination four weeks after the second dose for immunocompromised persons [1]. A fourth vaccination is recommended for vulnerable people [2].

mRNA vaccines against SARS-CoV-2 have been shown to be safe in vulnerable, immunocompromised patient populations, e.g., oncologic patients. Accordingly, a large cohort study including 74,878 patients with active cancer or a history of cancer has found a low rate of vaccination-related adverse events [3]. Regarding the effectiveness of SARS-CoV-2 vaccinations, analyses of immune responses in vulnerable patients suggest relevant B- and T-cell reactivity [4]. However, seroconversion rates after the first vaccination have been found to be far lower than in healthy controls but increased after the second vaccination [5].

In line with these findings, studies on SARS-CoV-2 vaccination in immunocompromised transplant recipients suggest an impaired response towards SARS-CoV-2 initial vaccination with increasing serologic responses after three or more doses of vaccine [6]. Patients with multiple sclerosis (MS) also belong to the group of vulnerable people as treating their disease requires immunomodulating disease-modifying therapies (DMTs).

Ofatumumab, a human anti-CD20 monoclonal antibody for monthly subcutaneous application, is approved in Europe for active relapsing MS (RMS) in adults. Its mode of action involves the selective depletion of CD20+ B-cells but spares CD20-negative long-lived plasma cells. While the first are major contributors to the adaptive immune system, the latter are especially important for immune memory [7].

For B-cell depleting therapies other than ofatumumab, it is recommended to wait at least three to six months after the last injection before vaccines are applied. This is not very compatible with the recommended regimen of SARS-CoV-2 mRNA vaccination. The interruption of treatment or delayed initiation of treatment should be avoided because of the risk of disease progression [8,9]; however, it is important to assess whether vaccination under continuous ofatumumab therapy elicits an immune response.

According to a retrospective chart review, seroconversion for SARS-CoV-2 neutralizing antibodies is impaired in MS patients receiving anti-CD20 antibodies, with lower seroconversion rates with rituximab and ocrelizumab compared to ofatumumab [10]. In line with these retrospective results, it has recently been shown in a prospective setting that ofatumumab-treated patients respond to initial vaccination against SARS-CoV-2 and while neutralizing antibody titers were reduced under ofatumumab, T-cell response was not affected [11,12]. SARS-CoV-2 infections after two doses of vaccine in patients receiving anti-CD20 antibodies were reported to be mild or asymptomatic and almost all patients seroconverted after infection [13]. Similarly, it has been suggested that booster vaccinations are of particular relevance for increasing antibody titers [14]. Regarding T-cell responses, conflicting findings either suggest no significant effect on T-cell levels [14] or the induction of memory T-cell levels after booster vaccinations [15]. Moreover, there are still few data on how patients with MS receiving treatment with ofatumumab respond to booster vaccination. Some data collected in patients receiving injectable anti-CD20 antibodies suggest a significant increase in antibody titers after a third SARS-CoV-2 vaccination [16]. Similar results were found in an observational study including patients with ofatumumab treatment. In that study, the third vaccination increased IgG titers by a factor of 1.4 to 1.6 compared to titers after the second vaccination. This study also suggests that after three vaccinations, during ofatumumab treatment, B-cell and T-cell responses were still lower than among healthy controls and compared to patients who received two or three doses of vaccine before ofatumumab was started [17].

The objective of the KYRIOS study was to examine the immune response after the completion of initial vaccination against SARS-CoV-2 with mRNA vaccines as well as the immune response to a booster SARS-CoV-2 mRNA vaccine. Therefore, the presence of SARS-CoV-2-specific T-cells and neutralizing as well as total antibodies was analyzed in the KYRIOS study. We here present the results in the subpopulation of patients who received a booster vaccination. The results on the initial vaccination population have been reported previously [11].

2. Materials and Methods

Details on study design, participants, treatments, outcomes and assessments as well as statistical analyses have been published previously [11]. Briefly, KYRIOS is a prospective, open-label, multicenter study (EudraCT 2021-000307-20; NCT04869358) designed to investigate the immune response towards SARS-CoV-2 mRNA vaccines in RMS patients in whom ofatumumab has already been initiated or who are planned to be initiated on ofatumumab upon the physician's discretion.

The original study protocol comprised cohort 1 and cohort 2, which included patients who received their initial vaccination (i.e., the first and second doses of SARS-CoV-2 mRNA vaccines) during the study either before ofatumumab was started or during stable ofatumumab therapy (Figure 1). Stable ofatumumab treatment was defined as treatment that had been started at least four weeks ago. The depletion of B-cells was verified before vaccination. The week 1 and month 1 results of cohort 1 and cohort 2 have been published previously [11]. Booster vaccinations in these cohorts were optional. We here report the booster vaccination results of 2 patients in cohort 2 (Figure 1).

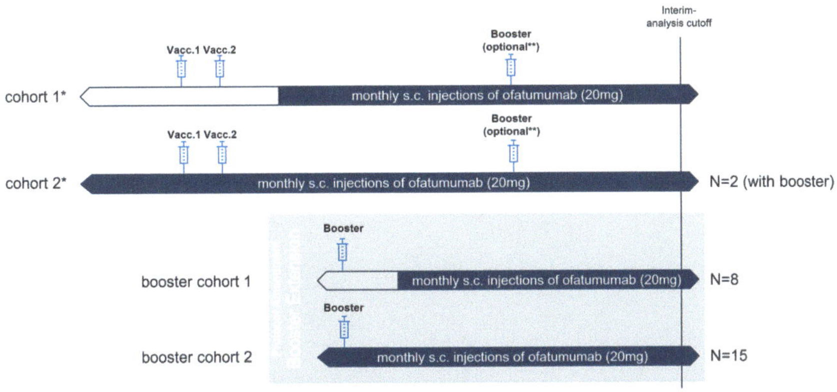

Figure 1. Description of different cohorts in KYRIOS. Cohort 1 and cohort 2 form the initial vaccination cohorts, i.e., patients who received the first and second dose of the SARS-CoV-2 vaccine during the KYRIOS study, either before or during ofatumumab treatment. Two patients of cohort 2 have already received a booster vaccination within the KYRIOS study. Booster cohort 1 and booster cohort 2 consist of patients who received only their booster vaccination during the KYRIOS study (but not the initial vaccination), either before or during ofatumumab treatment. Month 1 results after booster vaccination are reported for booster cohort 1, booster cohort 2, and for the two patients of the initial cohort 2. * Week 1 and month 1 results after the initial vaccination in cohort 1 (N = 6) and cohort 2 (N = 5) have been reported previously [11]. ** Booster vaccinations were optional in cohort 1 and cohort 2 and can be performed any time after the second dose of SARS-CoV-2 vaccine.

A protocol amendment introduced booster cohorts 1 and 2 to the study. These cohorts included patients who had already completed the initial vaccination cycle outside the study and received only their booster vaccine during the study (Figure 1). Booster cohort 2 patients may have received their initial vaccination before or during ofatumumab treatment. We here report results of booster cohort 1 and 2.

An additional cohort (cohort 3) with patients from the AMA-VACC study was included as the historical control group. These patients were booster vaccinated while receiving treatment with dimethyl fumarate (DMF), glatiramer acetate (GA), beta-interferons (IFN), or teriflunomide (TF) or while not being treated with a DMT [18].

The primary endpoint was the proportion of patients with SARS-CoV-2-specific T-cell response one month after completion of the booster vaccination. T-cell response was defined as the presence of SARS-CoV-2-reactive T-cells secreting either IFN-γ or IL-2 or both (any T-cell activity above the basal level). Secondary endpoints included the following: the extent of T-cell response defined as T-cell reactivity normalized for basal T-cell activity measured by IFN-γ secretion (IFN-γ stimulation indices); the proportion of patients with neutralizing antibodies against SARS-CoV-2; titers of serum total and neutralizing antibodies against SARS-CoV-2; the incidence of COVID-19 after complete vaccination; and comparison of immune responses in KYRIOS cohorts with the responses in cohort 3 derived from AMA-VACC.

Assessment time points for the booster cohorts in the KYRIOS study were month 1, month 6, month 12 and month 18. We present a pre-planned interim analysis of data obtained one month after the booster vaccination (data cut-off: 12 July 2022). The study is ongoing and follow-up data collected at further time points will be presented upon study completion. All endpoints were analyzed descriptively without formal statistical testing.

The study is consistent with the Declaration of Helsinki and conducted according to the Good Clinical Practice guidelines by the International Conference on Harmonisation (ICH-GCP). Ethics committee approval was obtained, and all patients or their legal representatives provided written informed consent before any study-related procedures were started.

Of note, recruitment into this study started in May 2021, which was before the Robert Koch-Institute issued its recommendation for a third vaccination as soon as four weeks after the second vaccination in severely immunocompromised patients in September 2021 [1]. Only in December 2021 was a joint statement published by the Deutsche Multiple Sklerose Gesellschaft (DMSG), the Kompetenznetz Multiple Sklerose (KKNMS) and the Berufsverband Deutscher Neurologen (BDN) which specified that this recommendation should be applied to MS patients receiving anti-CD20 antibodies or S1P-inhibitors [19].

3. Results

The KYRIOS study included 23 patients who received their initial vaccination (first and second dose) outside the study and their booster vaccination during the study. All these patients were initially vaccinated before ofatumumab treatment was started; eight patients also had their booster vaccine before starting ofatumumab (booster cohort 1) while fifteen patients received their booster while being continuously treated with ofatumumab (booster cohort 2). Furthermore, data from two patients in the initial cohort 2 are available, who received their initial and their booster vaccination during the study, both while being treated with ofatumumab.

Table 1 shows the patient characteristics of the booster population at screening. Briefly, patients in booster cohort 1 were on average 47.1 years of age and 45.5 years of age in booster cohort 2. The disease was diagnosed on average 11.1 and 7.2 years ago, respectively. In total, 37.5% and 33.3% of patients were previously untreated in booster cohort 1 and booster cohort 2, respectively. Mostly, mRNA vaccines by BioNTech/Pfizer were administered as the booster, with an average of 26 weeks after the second dose both in booster cohort 1 and booster cohort 2. The mean time between booster vaccination and the start of ofatumumab in booster cohort 1 was 0.87 months. In booster cohort 2, ofatumumab was started on average 1.87 months before booster vaccination (Table 2). Cohort 3 included 20 patients from the AMA-VACC study treated with DMF, GA, IFN or TF, of whom 18 patients received a booster vaccination. The patient characteristics of the AMA-VACC population are presented in Table 1, as published previously [18]. In total, 62.5% of booster cohort 1 and 66.7% of booster cohort 2 had received DMTs prior to ofatumumab. In most cases, this treatment was continued during initial vaccination. Overall, 37.5% patients in booster cohort 1 and 46.7% patients in booster cohort 2 received DMTs during their initial vaccination.

Table 1. Demographic and disease characteristics.

Variable *	Booster Cohort 1 Vaccination Prior to Treatment	Booster Cohort 2 Vaccination during Stable Treatment	Cohort 3 Vaccination during IFN/GA/DMF/TF/ no DMT
N	8	15	20 [a]
Age, years	47.1 (14.1)	45.5 (12.4)	48.6 (12.9)
Sex, female, n (%)	5 (62.5)	9 (60.0)	16 (80.0)
Time since diagnosis, years	11.1 (8.7)	7.2 (7.7)	13.99 (10.43)
Number of prior DMTs	2.0 (2.1)	1.3 (1.2)	1.6 (0.7)

Table 1. *Cont.*

Variable *	Booster Cohort 1 Vaccination Prior to Treatment	Booster Cohort 2 Vaccination during Stable Treatment	Cohort 3 Vaccination during IFN/GA/DMF/TF/ no DMT
Number of DMTs prior to ofatumumab, n (%)			
0	3 (37.5)	5 (33.3)	0 (0.0)
≥1	5 (62.5)	10 (66.7)	18 (100.00)

* If not indicated otherwise, data are presented as mean (SD). [a]: 18 of 20 patients in the AMA-VACC study had a booster vaccination. DMF: dimethyl fumarate; GA: glatiramer acetate, IFN: interferon-beta; TF: teriflunomide.

Table 2. Vaccination characteristics.

Variable	Booster Cohort 1 Vaccination Prior to Treatment	Booster Cohort 2 Vaccination during Stable Treatment	Cohort 3 Vaccination during IFN/GA/DMF/TF/ no DMT
N	8	15	20 [a]
Vaccination, n (%)			
1st (BioNTech/Pfizer ǀ Moderna)	8 (100.0) ǀ 0 (0.0)	14 (93.3) ǀ 1 (6.7)	19 (95.0) ǀ 1 (5.0)
2nd (BioNTech/Pfizer ǀ Moderna)	8 (100.0) ǀ 0 (0.0)	14 (93.3) ǀ 1 (6.7)	19 (95.0) ǀ 1 (5.0)
First booster vaccination (BioNTech/Pfizer ǀ Moderna)	7 (87.5) ǀ 1 (7.1)	13 (65.0) ǀ 2 (10.0)	11 (61.1) ǀ 7 (38.9)
Vaccination time interval, mean (SD)			
1st to 2nd vaccination, weeks/days	5.6 (0.7) weeks	5.6 (1.4) weeks	36.8 (9.0) days
2nd vaccination to booster vaccination, weeks/months	26.0 (2.3) weeks	26.2 (5.9) weeks	5.82 (0.4) months
Time interval between start of ofatumumab and vaccination, mean (SD)			
Booster vaccination to start of ofatumumab, months	0.87 (0.18)	-	-
Start of ofatumumab to booster vaccination, months	-	1.87 (0.85)	-

[a]: 18 of 20 patients in the AMA-VACC study had a booster vaccination. DMF: dimethyl fumarate; GA: glatiramer acetate, IFN: interferon-beta; TF: teriflunomide.

We observed a T-cell response according to the primary endpoint definition (presence of SARS-CoV-2 reactive T-cells secreting either IFN-γ or IL-2 or both) in 87.5% of patients in booster cohort 1 and in 46.7% of patients in booster cohort 2. The extent of T-cell response (T-cell reactivity measured by IFN-γ secretion normalized for basal T-cell activity, i.e., IFN-γ stimulation indices) was not markedly different between the two booster cohorts (Figure 2).

Neutralizing antibody titers increased in booster cohort 1 and 2. The seroconversion rate in booster cohort 1 was 87.5% prior to and 100.0% after the booster in booster cohort 1, and increased from 71.4% prior to the booster to 93.3% at month 1 after the booster in booster cohort 2. Patients in booster cohort 2 with prevalent antibodies before the booster reached similar titers as booster cohort 1 and 3. Of note, of the four patients of booster cohort 2 who were seronegative after the initial vaccination, three patients reached seroconversion after booster vaccination. Of the two patients who had both their initial and booster vaccination during the study while receiving ofatumumab treatment, one patient was seropositive for neutralizing antibodies already prior to the booster and showed an increase in antibody titers at month 1. The second patient was seronegative before and seroconverted after the booster vaccination. In cohort 3, all but two patients were seropositive before the booster. Both patients were seropositive after the booster vaccination (Figure 3A).

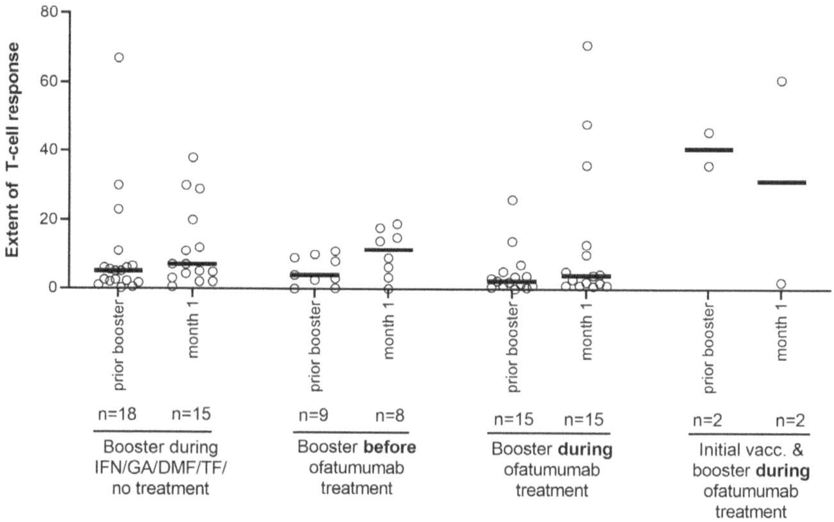

Figure 2. ELISpot-based quantification of T-cell reactivity after booster vaccination prior to or during ofatumumab treatment by calculation of IFN-γ stimulation indices towards SARS-CoV-2. Each dot represents one patient; medians are indicated by horizontal lines. All patients received their initial vaccination cycle before starting ofatumumab treatment (except for 2 patients with initial and booster during ofatumumab treatment). DMF: dimethyl fumarate; GA: glatiramer acetate, IFN: interferon-beta; n: number of patients with assessments; TF: teriflunomide.

Before the booster vaccination, all but three patients (one in booster cohort 1; two in cohort 3) had SARS-CoV-2 serum total antibody titers of 250 U/mL (assay-specific maximum of quantification range). After the booster vaccination, all patients in both booster cohorts had reached titers of 250 U/mL. In booster cohort 2, all but four patients had SARS-CoV-2 serum total antibody titers of 250 U/mL before the booster. In all these patients, total anti-spike antibody titers increased after booster vaccination. Of note, one patient in booster cohort 2 was seronegative for total antibodies before and seroconverted after the booster vaccination. Furthermore, of the two patients with initial and booster vaccination during the study while receiving ofatumumab, both reached serum total antibody titers of 250 U/mL after their booster (Figure 3B).

The median time of observation in the study until the data cut-off was 30.9 weeks. During this period, adverse events (AEs) were reported in 19 patients (82.6%), with 7 cases being related to the DMT and 2 cases being related to the vaccine, i.e., fatigue and tinnitus, which have both been reported as mild (Table 3). Only one relapse in one patient occurred (booster cohort 1). No serious AEs and no deaths were observed.

During the observational period, six cases of clinical COVID-19 were reported, one in booster cohort 1 and five in booster cohort 2. The infections occurred two months after the second vaccination in booster cohort 1, and two months (one patient), three months (one patient), as well as four months (three patients) after the second vaccination in booster cohort 2. All infections were mild or moderate according to CTCAE grading with full recovery in all cases. In booster cohort 1, the infection lasted 7 days, while the duration of infection in cohort 2 ranged from 8 to 14 days.

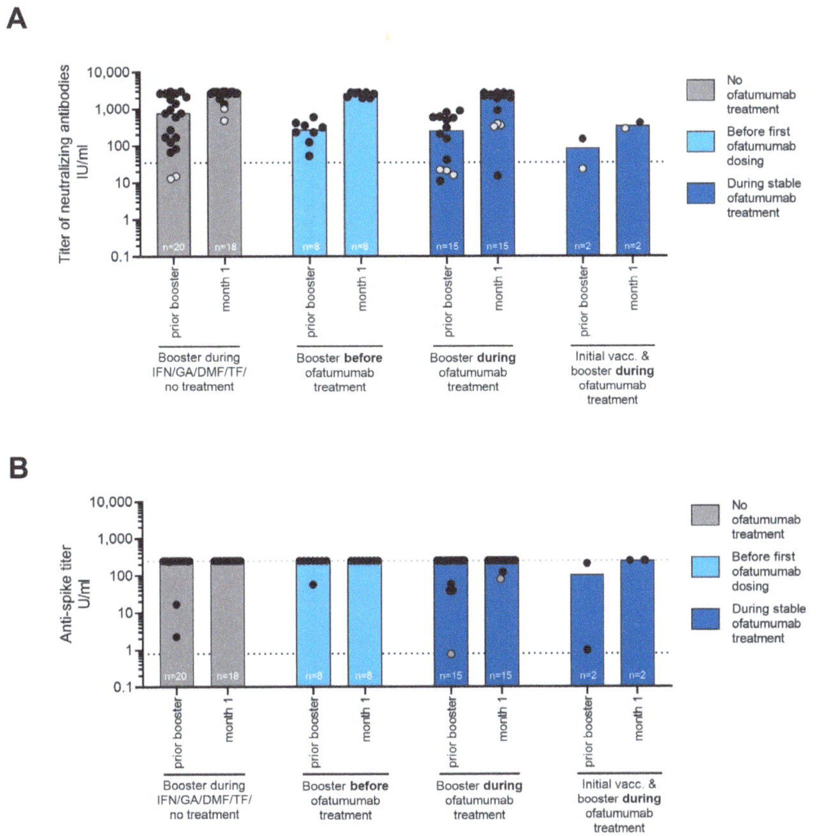

Figure 3. (**A**) Quantification of SARS-CoV-2-specific neutralizing antibody titer in U/mL after booster vaccination prior to or during ofatumumab treatment. (**B**) SARS-CoV-2-specific serum total antibody titer in U/mL after booster vaccination prior to or during ofatumumab treatment. All patients with available data were included in the analysis and individual values are represented by dots. Grey dots = patients who seroconverted after booster. Bars show median values, black dotted lines indicate assay-specific cut-off for seropositivity and grey dotted line indicates the maximal value of quantification range. DMF: dimethyl fumarate; GA: glatiramer acetate, IFN: interferon-beta; n: number of patients with assessments; TF: teriflunomide.

Table 3. Overview of adverse events.

Adverse Events, n (%)	Booster Cohort 1 Booster Vaccination Prior to Treatment (N = 8)	Booster Cohort 2 Booster Vaccination during Stable Treatment (N = 15)
Adverse events (AEs)	6 (75.0)	13 (86.7)
General disorders and administration site conditions	1 (12.5)	3 (20.0)
Nervous system disorders	3 (37.5)	4 (26.7)
Musculoskeletal and connective tissue disorders	2 (25.0)	2 (13.3)
Blood and lymphatic system disorders	0 (0.0)	0 (0.0)
Infections and infestations	3 (37.5)	9 (60.0)
Ear and labyrinth disorders	0 (0.0)	1 (6.7)

Table 3. *Cont.*

Adverse Events, n (%)	Booster Cohort 1 Booster Vaccination Prior to Treatment (N = 8)	Booster Cohort 2 Booster Vaccination during Stable Treatment (N = 15)
Gastrointestinal disorders	1 (12.5)	0 (0.0)
Injury, poisoning and procedural complications	2 (25.0)	1 (6.7)
Metabolism and nutrition disorders	1 (12.5)	0 (0.0)
Psychiatric disorders	0 (0.0)	1 (6.7)
Reproductive system and breast disorders	2 (25.0)	0 (0.0)
Respiratory, thoracic and mediastinal disorders	1 (12.5)	0 (0.0)
Skin and subcutaneous tissue disorders	1 (12.5)	1 (6.7)
Vascular disorders	0 (0.0)	2 (13.3)
Not coded	2 (25.0)	2 (13.3)
AEs related to DMTs	4 (50.0)	3 (20.0)
AEs related to SARS-CoV-2 vaccine	0 (0.0)	2 (13.3)
AEs leading to permanent discontinuation of study medication	0 (0.0)	0 (0.0)
AEs leading to temporary interruption of study medication	0 (0.0)	1 (6.7)
Serious adverse events	0 (0.0)	0 (0.0)

In case of multiple AEs, a patient is counted only once in the respective category.

4. Discussion

We here report the results on the subpopulation of patients who received a booster vaccination in the KYRIOS study. The results on the initial vaccination population have been published previously [11]. The present results therefore add to the previous results and provide essential insights on the immune response to booster vaccinations before and during ofatumumab therapy.

Accordingly, T-cell reactivity towards SARS-CoV-2 vaccines showed a slight increase in all booster cohorts at month 1 after the booster vaccination, irrespective of whether patients were vaccinated prior to ofatumumab or during continuous ofatumumab or whether they received other DMTs when vaccinated.

As far as T-cell response is concerned, it has been previously shown by several analyses that anti-CD20 treatment including ofatumumab does not alter T-cell reactivity towards SARS-CoV-2 [11,12,20]. The fact that T-cell reactivity only showed a minor increase in our analysis of booster vaccinations might be an issue of timing of assessments. T-cell reactivity becomes difficult to detect at month 1, as T-cells specific for SARS-CoV-2 then no longer circulate in the blood and tissue-resident memory T-cells develop [21]. It can be assumed that in patients receiving a booster vaccination, T-cell reactivity peaks even earlier due to altered immune mechanisms in immunized patients. However, as antibodies were detected, prior T-cell reactivity in the booster cohorts can be assumed, as T-cells are essential for B-cell-mediated antibody formation in response to mRNA vaccines [22]. It has to be noted that booster vaccinations are of special relevance for increasing antibody levels [14].

Booster vaccinations showed a similar humoral response irrespective of whether the booster was applied prior to or during ofatumumab treatment. All patients who had their booster vaccination during ofatumumab treatment showed an increase in neutralizing antibodies comparable to the cohorts vaccinated prior to ofatumumab or under other DMTs. However, while the relative increase was higher in patients with lower levels before the booster, higher pre-booster antibody levels seem to be associated with a higher absolute antibody level after the booster [23]. Antibody levels prior to the booster, in turn, are likely to depend on prior treatment. In the study by Faissner et al., an impaired humoral response was reported [12], but baseline antibody levels have not been assessed. However, three patients were treated with sphingosine-1-phosphate (S1P) inhibitors during their initial

SARS-CoV-2 vaccination (i.e., first and second dose of the vaccine) [12]. S1P inhibitors are known to be associated with reduced antibody levels in response to vaccination during continuous treatment [18]. Therefore, prior treatments in the Faissner study might have impacted the antibody development after vaccination against SARS-CoV-2. In the KYRIOS study, seronegative patients or patients with lower titers before the booster vaccination showed almost similar increases, and 75% of seronegative patients before the booster vaccination reached seropositivity at month 1 even when vaccinated during continuous ofatumumab. Of note, two-thirds of the patients in booster cohort 2 had received other DMTs prior to ofatumumab, including S1P-inhibitors. It seems reasonable to assume that these prior treatments might have also affected the immune response to SARS-CoV-2 vaccination, especially when DMTs were switched only shortly before the vaccination.

The fact that booster vaccination further increased neutralizing antibody titers also in those who received their first vaccination during ofatumumab suggests the development of immune memory after their initial vaccination. The application of a booster vaccination then activates humoral immune memory and leads to the proliferation of antigen-specific B-cells. Preclinical data on ofatumumab support these findings. In contrast to ocrelizumab, human equivalent doses of ofatumumab applied in huCD20 transgenic mice were shown to spare marginal zone and follicular B cells in lymphoid organs and in the bone marrow, which are important for immune surveillance, B-cell repletion and preservation of the immune response [24].

Regarding booster vaccinations, the KYRIOS study suggested similar immune responses in patients receiving the booster either prior to or during ofatumumab. It can be assumed that it is not necessary to postpone ofatumumab treatment initiation until after the booster vaccination. Furthermore, booster data indicate that patients who had received their initial vaccination during ofatumumab benefit from an additional booster vaccination. This supports the vaccination recommendations toward a third vaccination as soon as four weeks after the second vaccination in MS patients receiving anti-CD20 antibodies or S1P-inhibitors [1,19]. It can be assumed that as in immunosuppressed transplant recipients, each booster leads to an increase in seroconversion rate [6].

In general, higher antibody titers are correlated with better protection from severe COVID-19. The KYRIOS results suggest clinically effective vaccinations in MS patients receiving ofatumumab, irrespective of whether vaccination is completed before treatment initiation or whether it is applied during ofatumumab treatment, as no severe infections occurred, and infections only lasted 7 to 14 days. Therefore, the severity and duration of COVID-19 in KYRIOS are well in line with ALITHIOS study results on infection in ofatumumab-treated patients [25]. However, as we have previously pointed out [11], COVID-19 cases in ALITHIOS were observed before September 2021, and therefore occurred before the circulation of Omicron. On the contrary, COVID-19 cases in KYRIOS mainly occurred during early 2022. It can be assumed that a relevant number of these cases were Omicron infections. Omicron is known to escape the immune response, but still leaves vaccines effective in preventing severe COVID-19. According to the KYRIOS study, the latter also applies to patients receiving ofatumumab. Given this background and our present data, a booster with the Omicron-adapted vaccine seems reasonable.

SARS-CoV-2 mRNA vaccines were well tolerated in the present study, with only two cases of mild adverse events related to the booster vaccination being reported. No thrombosis or cardiovascular events have been reported in the booster cohorts. Vaccine-induced immune thrombocytopenia and thrombosis (VITT) are a rare but serious issue associated with SARS-CoV-2 mRNA vaccination [26], possibly mediated by anti-PF4 antibodies, which can be found after vaccination but not after COVID-19 infection [27].

When interpreting the results of KYRIOS, some study limitations have to be kept in mind. As KYRIOS only included a small sample, the results should not be overinterpreted as they require further confirmation. As no week 1 assessment was scheduled for patients who received a booster vaccination, information on early T-cell responses after booster vaccination is lacking. Furthermore, patients were included in the study before the relevant authorities issued a recommendation to apply a third dose as early as four weeks after the second. The time between the first and second dose of the vaccination cycle and booster vaccination was therefore longer (median 6 months) in our study than what is currently recommended.

5. Conclusions

Overall, the present KYRIOS data add to the previous results on the immune response towards SARS-CoV-2 mRNA vaccination. The previously reported results gave valuable insights into the impact of concurrent ofatumumab treatment during the initial vaccination cycle. Given that booster vaccinations are generally recommended, information on immune responses to booster vaccinations before and during ofatumumab therapy were needed. The novel KYRIOS booster data show that a booster vaccination increases neutralizing antibody titers in patients continuing ofatumumab therapy. A similar antibody titer is achieved as that in patients not undergoing ofatumumab therapy. The present data give the first indications that a booster may possibly lead to seroconversion in patients who were seronegative after initial vaccination and received a booster during stable ofatumumab treatment, regardless of which therapy the patients had received during the initial vaccination, including ofatumumab. Further data are needed to support this assumption. Since higher titers correlate with vaccine effectiveness, a booster is recommended in patients on ofatumumab therapy.

Author Contributions: T.Z.: conceptualization, investigation and writing (review and editing); E.S.: investigation and writing (review and editing); M.G.: investigation, project administration and writing (review and editing); B.E.: conceptualization, project administration, investigation and writing (review and editing); T.B.: conceptualization, investigation and writing (review and editing). All authors have read and agreed to the published version of the manuscript.

Funding: This study and medical writing support were sponsored by Novartis Pharma Vertriebs GmbH.

Institutional Review Board Statement: The study was conducted in accordance with the guidelines of the Declaration of Helsinki, and approved by the Ethics Committee of the Technische Universität Dresden (protocol code COMB157GDE01, date of approval 28 April 2021).

Informed Consent Statement: Informed consent was obtained from all subjects involved in the study.

Data Availability Statement: Data will be provided upon reasonable request.

Acknowledgments: Medical writing support (preparation of the manuscript draft) was provided by Karin Eichele (mediwiz).

Conflicts of Interest: Tjalf Ziemssen has received personal compensation for participating in advisory boards, trial steering committees, and data and safety monitoring committees, as well as for scientific talks and project support from Almirall, Bayer, BAT, Biogen, Celgene, Sanofi Genzyme, Merck, Novartis, Roche, Vitaccess and Teva. Eugen Schlegel has received research support, consulting fees and honoraria for lectures from Biogen, Lilly, Merck and Novartis. Marie Groth and Benjamin Ettle are employees of Novartis Pharma GmbH, Nuremberg, Germany. Tobias Bopp has received consulting fees and honoraria for lectures from Biogen, Celgene, Merck, Novartis, Pathios Therapeutics, Roche, Sanofi Genzyme and Teva.

References

1. Robert Koch-Institut. Epidemiologisches. In *Bulletin*; Robert Koch-Institut: Berlin, Germany, 2021.
2. Robert Koch-Institut. Epidemiologisches. In *Bulletin*; Robert Koch-Institut: Berlin, Germany, 2022.
3. Kang, W.; Shami, J.J.P.; Yan, V.K.C.; Ye, X.; Blais, J.E.; Li, X.; Lee, V.H.F.; Chui, C.S.L.; Lai, F.T.T.; Wan, E.Y.F.; et al. Safety of two-dose COVID-19 vaccination (BNT162b2 and CoronaVac) in adults with cancer: A territory-wide cohort study. *J. Hematol. Oncol.* **2022**, *15*, 66. [CrossRef]
4. Echaide, M.; Labiano, I.; Delgado, M.; Fernández de Lascoiti, A.; Ochoa, P.; Garnica, M.; Ramos, P.; Chocarro, L.; Fernández, L.; Arasanz, H.; et al. Immune Profiling Uncovers Memory T-Cell Responses with a Th17 Signature in Cancer Patients with Previous SARS-CoV-2 Infection Followed by mRNA Vaccination. *Cancers* **2022**, *14*, 4464. [CrossRef] [PubMed]

5. Goshen-Lago, T.; Waldhorn, I.; Holland, R.; Szwarcwort-Cohen, M.; Reiner-Benaim, A.; Shachor-Meyouhas, Y.; Hussein, K.; Fahoum, L.; Baruch, M.; Peer, A.; et al. Serologic Status and Toxic Effects of the SARS-CoV-2 BNT162b2 Vaccine in Patients Undergoing Treatment for Cancer. *JAMA Oncol.* **2021**, *7*, 1507–1513. [CrossRef] [PubMed]
6. Osmanodja, B.; Ronicke, S.; Budde, K.; Jens, A.; Hammett, C.; Koch, N.; Seelow, E.; Waiser, J.; Zukunft, B.; Bachmann, F.; et al. Serological Response to Three, Four and Five Doses of SARS-CoV-2 Vaccine in Kidney Transplant Recipients. *J. Clin. Med.* **2022**, *11*, 2565. [CrossRef] [PubMed]
7. Lee, D.S.W.; Rojas, O.L.; Gommerman, J.L. B cell depletion therapies in autoimmune disease: Advances and mechanistic insights. *Nat. Rev. Drug Discov.* **2021**, *20*, 179–199. [CrossRef]
8. He, A.; Merkel, B.; Brown, J.W.L.; Zhovits Ryerson, L.; Kister, I.; Malpas, C.B.; Sharmin, S.; Horakova, D.; Kubala Havrdova, E.; Spelman, T.; et al. Timing of high-efficacy therapy for multiple sclerosis: A retrospective observational cohort study. *Lancet Neurol.* **2020**, *19*, 307–316. [CrossRef]
9. Koeditz, D.; Frensch, J.; Bierbaum, M.; Ness, N.H.; Ettle, B.; Vudumula, U.; Gudala, K.; Adlard, N.; Tiwari, S.; Ziemssen, T. Comparing the long-term clinical and economic impact of ofatumumab versus dimethyl fumarate and glatiramer acetate in patients with relapsing multiple sclerosis: A cost-consequence analysis from a societal perspective in Germany. *Mult. Scler. J. Exp. Transl. Clin.* **2022**, *8*, 20552173221085741. [CrossRef]
10. Levit, E.; Longbrake, E.E.; Stoll, S.S. Seroconversion after COVID-19 vaccination for multiple sclerosis patients on high efficacy disease modifying medications. *Mult. Scler. Relat. Disord.* **2022**, *60*, 103719. [CrossRef]
11. Ziemssen, T.; Groth, M.; Ettle, B.; Bopp, T. Immune Response to SARS-CoV-2 mRNA Vaccines in an Open-Label Multicenter Study in Participants with Relapsing Multiple Sclerosis Treated with Ofatumumab. *Vaccines* **2022**, *10*, 2167. [CrossRef]
12. Faissner, S.; Heitmann, N.; Plaza-Sirvent, C.; Trendelenburg, P.; Ceylan, U.; Motte, J.; Bessen, C.; Urlaub, D.; Watzl, C.; Overheu, O.; et al. Immune response in ofatumumab treated multiple sclerosis patients after SARS-CoV-2 vaccination. *Front. Immunol.* **2022**, *13*, 980526. [CrossRef]
13. Stoll, S.; Desai, S.; Levit, E. A retrospective evaluation of seroconversion after COVID-19 during the early Omicron wave in fully vaccinated multiple sclerosis patients receiving anti-CD20 therapies. *Mult. Scler. Relat. Disord.* **2023**, *71*, 104574. [CrossRef]
14. Maringer, Y.; Nelde, A.; Schroeder, S.M.; Schuhmacher, J.; Hörber, S.; Peter, A.; Karbach, J.; Jäger, E.; Walz, J.S. Durable spike-specific T cell responses after different COVID-19 vaccination regimens are not further enhanced by booster vaccination. *Sci. Immunol.* **2022**, *7*, eadd3899. [CrossRef]
15. Mise-Omata, S.; Ikeda, M.; Takeshita, M.; Uwamino, Y.; Wakui, M.; Arai, T.; Yoshifuji, A.; Murano, K.; Siomi, H.; Nakagawara, K.; et al. Memory B Cells and Memory T Cells Induced by SARS-CoV-2 Booster Vaccination or Infection Show Different Dynamics and Responsiveness to the Omicron Variant. *J. Immunol.* **2022**, *209*, 2104–2113. [CrossRef] [PubMed]
16. König, M.; Torgauten, H.M.; Tran, T.T.; Holmøy, T.; Vaage, J.T.; Lund-Johansen, F.; Nygaard, G.O. Immunogenicity and Safety of a Third SARS-CoV-2 Vaccine Dose in Patients With Multiple Sclerosis and Weak Immune Response After COVID-19 Vaccination. *JAMA Neurol.* **2022**, *79*, 307–309. [CrossRef] [PubMed]
17. Sabatino, J.J., Jr.; Mittl, K.; Rowles, W.; Zamecnik, C.R.; Loudermilk, R.P.; Gerungan, C.; Spencer, C.M.; Sagan, S.A.; Alexander, J.; McPolin, K.; et al. Longitudinal adaptive immune responses following sequential SARS-CoV-2 vaccinations in MS patients on anti-CD20 therapies and sphingosine-1-phosphate receptor modulators. *Mult. Scler. Relat. Disord.* **2023**, *70*, 104484. [CrossRef]
18. Ziemssen, T.; Groth, M.; Rauser, B.; Bopp, T. Assessing the immune response to SARS-CoV-2 mRNA vaccines in siponimod-treated patients: A nonrandomized controlled clinical trial (AMA-VACC). *Ther. Adv. Neurol. Disord.* **2022**, *15*, 17562864221135305. [CrossRef] [PubMed]
19. DMSG; KKNMS. Stellungnahme zur 3. Impfung Gegen SARS-CoV-2 bei Menschen mit Multipler Sklerose (MS). Available online: https://www.dmsg.de/news/detailansicht/stellungnahme-zur-3-impfung-gegen-sars-cov2-bei-menschen-mit-multipler-sklerose-ms (accessed on 2 January 2023).
20. Wu, X.; Wang, L.; Shen, L.; Tang, K. Response of COVID-19 vaccination in multiple sclerosis patients following disease-modifying therapies: A meta-analysis. *EBioMedicine* **2022**, *81*, 104102. [CrossRef]
21. Reinscheid, M.; Luxenburger, H.; Karl, V.; Graeser, A.; Giese, S.; Ciminski, K.; Reeg, D.B.; Oberhardt, V.; Roehlen, N.; Lang-Meli, J.; et al. COVID-19 mRNA booster vaccine induces transient CD8+ T effector cell responses while conserving the memory pool for subsequent reactivation. *Nat. Commun.* **2022**, *13*, 4631. [CrossRef]
22. Toor, S.M.; Saleh, R.; Sasidharan Nair, V.; Taha, R.Z.; Elkord, E. T-cell responses and therapies against SARS-CoV-2 infection. *Immunology* **2021**, *162*, 30–43. [CrossRef]
23. Pani, A.; Romandini, A.; Schianchi, A.; Senatore, M.; Gagliardi, O.M.; Gazzaniga, G.; Agliardi, S.; Conti, T.; Schenardi, P.A.; Maggi, M.; et al. Antibody Response to COVID-19 Booster Vaccination in Healthcare Workers. *Front. Immunol.* **2022**, *13*, 872667. [CrossRef]
24. Bigaud, M. Comparative pharmacology of ofatumumab versus ocrelizumab in humanised-CD20 transgenic mice. P298. In Proceedings of the 38th Congress of the European Committee for Treatment and Research in Multiple Sclerosis (ECTRIMS), Amsterdam, The Netherlands, 26–28 October 2022.
25. Cross, A.H.; Delgado, S.; Habek, M.; Davydovskaya, M.; Ward, B.J.; Cree, B.A.C.; Totolyan, N.; Pingili, R.; Mancione, L.; Hu, X.; et al. COVID-19 Outcomes and Vaccination in People with Relapsing Multiple Sclerosis Treated with Ofatumumab. *Neurol. Ther.* **2022**, *11*, 741–758. [CrossRef] [PubMed]

26. Yasmin, F.; Najeeb, H.; Naeem, U.; Moeed, A.; Atif, A.R.; Asghar, M.S.; Nimri, N.; Saleem, M.; Bandyopadhyay, D.; Krittanawong, C.; et al. Adverse events following COVID-19 mRNA vaccines: A systematic review of cardiovascular complication, thrombosis, and thrombocytopenia. *Immun. Inflamm. Dis.* **2023**, *11*, e807. [CrossRef] [PubMed]
27. Piccin, A.; Mullin, B.; Brown, A.; Benson, G. The lack of anti-PF4 antibodies in convalescent plasma from COVID-19 infected blood donor. *Transfus. Med.* **2023**. [CrossRef] [PubMed]

vaccines MDPI

Article

Interactions between Severe Allergy and Anxiety in Anti-SARS-CoV-2 Vaccinees

Chiara Asperti [1,2,†], Giovanni Benanti [1,2,†], Giuseppe A. Ramirez [1,2], Marco Russo [2], Benedetta Vai [2,3], Barbara Bramé [4], Naomi Viapiana [1], Serena Nannipieri [1,2], Maria Bernadette Cilona [1,2], Martina Mazzetti [2,5], Simone Zuffada [2,5], Valentina Elisabetta Di Mattei [2,5], Francesco Benedetti [2,3], Lorenzo Dagna [1,2,‡] and Mona-Rita Yacoub [1,2,*,‡]

1 Unit of Immunology Rheumatology, Allergy and Rare Diseases, IRCCS Ospedale San Raffaele, 20132 Milan, Italy
2 Faculty of Medicine, Università Vita-Salute San Raffaele, 20132 Milan, Italy
3 Psychiatry and Clinical Psychobiology, Division of Neuroscience, IRCCS San Raffaele Scientific Institute, 20132 Milan, Italy
4 Nuovo Ospedale di Legnano, Uoc Medicina Generale, 20025 Legnano, Italy
5 Clinical and Health Psychology Unit, IRCCS San Raffaele Scientific Institute, 20132 Milan, Italy
* Correspondence: yacoub.monarita@hsr.it
† These authors contributed equally to this work.
‡ These authors contributed equally to this work.

Abstract: Severe drug allergy affects patient hesitancy to new treatments, posing unprecedented challenges to anti-SARS-CoV-2 vaccination campaigns. We aimed to analyze the psychological profile of vaccinees with a history of severe allergy in comparison to subjects with a milder allergy history. Patients attending a dedicated vaccination setting were administered an anonymized questionnaire including clinical data and the State-Trait Anxiety Inventory (STAI) scale (score range 20–80). Patients were also asked whether being in a protected setting affected their attitude toward vaccination. Data are expressed as median (interquartile range). We enrolled 116 patients (78% women), of whom 79% had a history of drug anaphylaxis. The median state anxiety score was 36.5 (30–47.2), while the trait anxiety score was 37 (32–48). State anxiety was higher in those with severe than mild allergy [39 (32–50) vs. 30 (25–37); $p < 0.001$], with the highest score found in a patient with previous drug anaphylaxis (42.5 [32–51.7]). More than 50% of patients reported that being in a protected setting had lowered their anxiety. Severe allergy is associated with a higher burden of situational anxiety in the setting of vaccination without affecting patient constitutional (trait) levels of anxiety. Vaccination in dedicated facilities might overcome issues related to hesitancy and improve patients' quality of life.

Keywords: vaccine; COVID-19; SARS-CoV-2; allergy; anxiety

Citation: Asperti, C.; Benanti, G.; Ramirez, G.A.; Russo, M.; Vai, B.; Bramé, B.; Viapiana, N.; Nannipieri, S.; Cilona, M.B.; Mazzetti, M.; et al. Interactions between Severe Allergy and Anxiety in Anti-SARS-CoV-2 Vaccinees. *Vaccines* **2022**, *10*, 2047. https://doi.org/10.3390/vaccines10122047

Academic Editor: Kay Choong See

Received: 29 October 2022
Accepted: 28 November 2022
Published: 30 November 2022

1. Introduction

Anaphylaxis and severe allergic reactions constitute life-threatening events occurring with an estimated incidence of 4–5 cases per 100,000 persons per year [1]. Severe allergy survivors also face long-term psychological sequelae affecting their quality of life [1–3]. These subjects often develop a generalized sense of insecurity and anxiety, and they tend to be wary about changes in their medication and the administration of new drugs [3].

This might pose a relevant challenge in the setting of the ongoing severe acute respiratory syndrome coronavirus 2 (SARS-CoV-2) pandemic, which has prompted mass exposure to a vast array of newly developed treatments and vaccines [4].

True systemic hypersensitivity against vaccine is rare. For the Pfizer/BioNTech BNT162B2 COVID-19 vaccine, it occurs in 11.1 cases per 1 million doses [5]. Furthermore, 80% of patients with hypersensitivity reactions (HRs) to vaccines had a history of positive allergic reactions to food, drugs, or insect sting [6]. According to the European

Academy of Allergy and Clinical Immunology (EAACI), only patients with an established allergy to vaccine components have an absolute contraindication to vaccination [7]. Still, a relevant number of people have been inappropriately considered at-risk, and allergists have been insofar fundamental in assessing and identifying actual at-risk subjects [8]. As an example, in Hong Kong, allergist-led vaccination sessions had a vaccine recommendation rate of 98.9%, compared to 81% in the non-allergist-led one [9].

Nevertheless, people with a history of severe allergy should be vaccinated by staff able to recognize and treat allergic reactions [5]. In Italy, national guidelines recommend vaccination in a "protected setting" consisting in a medical center with dedicated staff and prolonged surveillance for these patients in order to protect their safety. Patients can receive indication directly from their allergist or can be referred by standard vaccination centers if deemed necessary, usually having a history of previous anaphylaxis, multiple drug reaction, previous suspected hypersensitivity reaction to COVID-19 vaccines (but negative skin tests to excipients), severe or uncontrolled asthma or chronic spontaneous urticaria (CSU) [10–12].

However, in the context of these "protected settings", allergists could also have a role in overcoming vaccine hesitancy by this special population [13].

In this study, we aimed to investigate anxiety levels of patients with a history of severe HR undergoing vaccination in "protected settings": we have compared state and trait anxiety levels between a Severe Allergic Group (SAG) and a group constituted by subjects with mild allergic background (Mild Allergic group = MAG). We also evaluated the potential effects of vaccination-protected setting on psychological well-being.

2. Materials and Methods

2.1. Patients and Procedures

We studied a cohort of consecutive allergic patients being referred to dedicated vaccination facilities at IRCCS San Raffaele Hospital (Milan, Italy) and Legnano Hospital (Legnano, Milan, Italy) from 8 October 2021 to 13 April 2022. Patients referred to these facilities had been deemed at risk for vaccine-related HRs [10] by either their allergist or a standard vaccination center. Enhanced safety measures included trained personnel for prompt resuscitation and prolonged post-vaccine observation for one hour. According to the provisions of our Hospital Institution, all subjects received the BNT162b2 (Comirnaty®) vaccine.

Based on clinical history, patients were subdivided first into SAG or MAG groups. SAG patients had a history of anaphylaxis or severe HR to drugs, foods, or insect stings (defined as grade two or higher of the word allergy organization classification [14]). MAG included patients with no severe allergic history (non-severe food allergy, well-controlled asthma or CSU, rhinitis/conjunctivitis, atopic dermatitis, and allergic contact dermatitis).

Next, we performed a second analysis by dividing our population into a Severe Drug Reaction (SDR) group and a non-SDR group in order to ascertain the specific role of drug allergy history.

Data collection was performed in the post-vaccine observation timeframe through an anonymous questionnaire. The questionnaire was designed in compliance with the European Commission guidelines for anonymization in such a way that patient identification was impossible for the investigators or other subjects [15]. For these reasons, the study did not require formal approval by the local Institutional Review Board. Collected data included patient demographics (gender, age range), general clinical history (comorbidities, ongoing therapies), and allergic history (previous severe reaction to vaccines or drugs, foods, insect stings, and respiratory or contact allergy). The number and type of previous anti-SARS-CoV-2 vaccines were also recorded. Anxiety was measured through the STAI-Y questionnaire. STAI-Y is a validated questionnaire, initially devised in 1970 and later revised in 1983 by Spielberger, one of the most used tools to analyze anxiety in medical research [16]. It provides a quantitative measurement of anxiety, separately analyzing the habitual proneness to anxiety (trait anxiety) and the in-the-moment anxiety to a specific event (state anxiety) [16]. Each section comprises 20 items, presented in both positive and negative forms, graded 1 to 4, with a total score ranging from 20 to 80. Higher scores are positively correlated with higher levels of anxiety; a cut-off score >39, as suggested by the literature, has been used to define clinically significant anxiety symptoms [17]. Patients were also asked whether having been referred to a dedicated facility had made them feel more or less anxious about vaccination.

2.2. *Statistical Analysis*

Shapiro-Wilk normality tests were performed to assess whether continuous variables had or not a normal distribution. Due to the non-normal distribution of continuous variables, non-parametric tests were employed. Correlation between continuous variables was performed with Spearman's test. Mann–Whitney's U-test was used to compare continuous variable trends between groups. The distribution of categorical variables among groups was compared using the Chi-square test with Fisher's exact correction. Continuous variables are expressed as median (interquartile range, IQR) unless otherwise specified. Categorical variables are reported as absolute numbers (percentages). RStudio 4.2.1. and JASP 0.16.0.0 were used for statistical analysis.

3. Results

SAG and MAG encompassed 89 and 27 subjects, respectively. In the SAG, 86% were women, and the most represented age range was 55–59 years. 78% of them reported previous drug anaphylaxis, and 56% food anaphylaxis. Allergic comorbidities (rhinitis, atopic dermatitis, asthma, pollen-fruit syndrome) were present in 67% of SAG and 56% of MAG. However, except for a history of food allergy, no significant differences were detected between SAG and MAG regarding the prevalence of allergic comorbidities. The demographic and clinical features of patients are shown in Table 1. SDR e non-SDR encompassed 69 and 47 patients, respectively. Demographic and clinical features of patients are shown in Table S1). Symptoms during post-vaccine observation were reported by 12.9%, but only 4.3% were suggestive of HR. In detail, four patients reported local pain in the injection site, four patients reported skin rash, one diffuse pruritus, one "oral itching", one headache, one heartburn, and one had a hypertensive episode. None of them reported systemic HRs or other any other severe adverse effect.

Table 1. Demographic and clinical features. Abbreviations: (HR) = Hypersensitivity reactions; (AD) = atopic dermatitis; (ACD) = allergic contact dermatitis; (CSU) = chronic spontaneous urticaria; (SPT) = Skin Prick Test; AntiH1 = antiH1 antihistamine; (STAI) = State-Trait Anxiety Inventory; (IQR) = Interquartile range, (NA) = not applicable.

	Total Sample	Severe Allergic Group	Mild Allergic Group	*p*
N (100%)	116 (100)	89 (77)	27 (23)	NA
Females: n (%)	90 (78)	76 (86)	15 (51)	<0.010
Age Class: median (IQR)	47 (37–57)	47 (37–57)	47 (32–57)	ns
Drug anaphylaxis: n (%)	70 (60)	70 (79)	0 (0)	<0.001
Drug HRs ≥ 2: n (%)	49 (42)	49 (56)	0 (0)	<0.001
Food anaphylaxis: n (%)	43 (37)	42 (48)	0 (0)	<0.001
Allergic comorbidities: n (%)	75 (66)	59 (67)	15 (56)	ns
Rhinitis/conjunctivitis: n (%)	37 (32)	30 (34)	7 (26)	ns
Asthma: n (%)	33 (28)	26 (30)	7 (26)	ns
AD: n (%)	33 (28)	27 (31)	5 (22)	ns
ACD: n (%)	29 (25)	27 (30)	2 (7)	<0.050
CSU: n (%)	12 (10)	9 (10)	3 (11)	ns
Food allergy: n (%)	47 (41)	42 (48)	5 (22)	<0.050
Hymenoptera allergy: n (%)	18 (16)	19 (22)	3 (11)	ns
Positive SPT: n (%)	60 (52)	51 (57)	9 (33)	<0.050
AntiH1 therapy: n (%)	49 (42)	39 (44)	10 (37)	ns
Inhaled asthma therapy	25 (21)	19 (21)	6 (22)	ns
Comorbidities: n (%)	19 (16)	16 (18)	3 (11)	ns

Table 1. *Cont.*

	Total Sample	Severe Allergic Group	Mild Allergic Group	*p*
Dose 1: n (%)	16 (14)	12 (14)	4 (15)	ns
Dose 2: n (%)	46 (40)	36 (41)	9 (33)	ns
Dose 3: n (%)	52 (45)	38 (43)	14 (52)	ns
Symptoms: n (%)	15 (13)	13 (15)	2 (7)	Ns
STAI-state: median (IQR)	36.5 (30.0–47.2)	39.0 (32.0–50.2)	30. 0 (24.5–36.5)	<0.050
STAI-trait: median (IQR)	37.0 (32.0–48.0)	37.50 (32.0–48.0)	37.0 (31.5–47.0)	Ns

Regarding the psychological impact of vaccine administration in a "protected setting", 60.3% answered that it made them feel less anxious, while only 9.4% were more anxious due to the hospital setting. Among the subgroup of patients with previous drug anaphylaxis, a significantly higher number of patients (71% in SDR vs. 50% in non-SDR, $p = 0.015$) reported that being in a "protected setting" made them feel less anxious (Figure S1).

Regarding the assessment of anxiety with the STAI-Y questionnaire, we found a statistically significant correlation between state anxiety and trait anxiety (rho = 0.580, $p = 0.001$). Gender did not correlate with a difference in anxiety level, while age range had a negative correlation with both state $p = 0.033$, rho = −0.200) and trait anxiety ($p = 0.030$, rho = −0.200), meaning that younger patients were more anxious than older patients both in general and during vaccination (Figures S2 and S3). SAG subjects had significantly greater post-vaccination state anxiety than subjects who reported no severe reactions in their history ($p < 0.001$). This trend was replicated in the SDR group ($p < 0.001$), where the difference between median state anxiety SDR and non-SDR (42.5 IQR [32–51.7] vs. 32.5 IQR [28–37.7]) was even greater than between SAG and MAG (39 [IQR 32–50] vs. 30 [IQR 24.5–36.5]). Moreover, both SAG and SDR groups had a median state anxiety level that was clinically significant. However, no significant differences in trait anxiety either between SAG and MAG or between SDR and the non-SDR group were found (Figure 1). Of note, subdividing patients according to the number of previous COVID-19 vaccinations yielded no significant differences in state or trait anxiety levels.

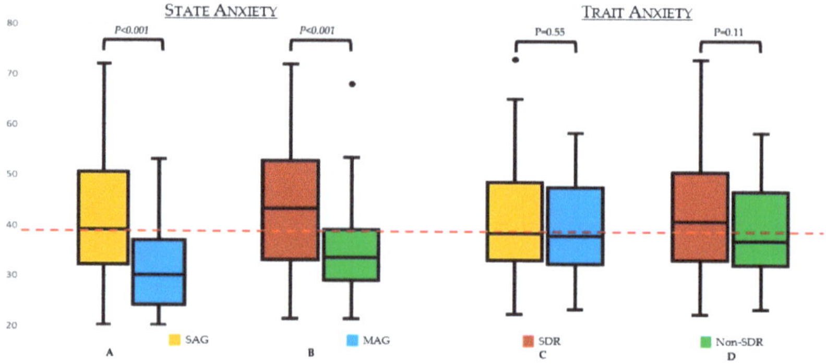

Figure 1. (**A**) State anxiety levels according to the STAY-Y tool in SAG and MAG. (**B**) State anxiety levels according to the STAY-Y tool in SDR and non-SDR groups. (**C**) Trait anxiety levels according to the STAY-Y tool in SAG and MAG. (**D**) Trait anxiety levels according to the STAY-Y tool in SDR and non-SDR groups. *p* values are shown on each of the graphs. The dotted line represents the clinical cut-off to define relevant anxiety level (>39). Using the STAY-Y tool, we assessed the anxiety level of subjects undergoing COVID-19 vaccination. First of all, we observed that patients with severe allergy background have a higher anxiety level during vaccination (39 IQR [32–50] vs. 30 IQR [24.5–36.4]), and the difference was even more evident in subjects with previous severe drug allergy (42.5 IQR [32–51.7] vs. 32.5 IQR [28–37.7]). On the contrary, no difference was observed in their trait anxiety (usual properness toward anxiety) between SAG and MAG (37.5 IQR [32.0–48.0] vs. 37 IQR [31.7–47)

or between the SDR group and the non-SDR group (40 IQR [32–49] vs. 36 IQR [31.5–46]). Of note, 39 is usually used as the cut-off to define a relevant state or trait anxiety, so it appears that MAG and non-SDR groups in the median did not have a relevant anxious state during vaccination, while both SAG and SDR groups presented with clinically relevant anxiety. SAG = severe allergy group; MAG = mild allergy group; SDR = severe drug reaction.

Other clinical features, such as atopy, CSU, and non-allergic comorbidities, were not associated with different levels of anxiety. Anxiety did not correlate with the onset of post-vaccination symptoms either in SAG or in MAG.

4. Discussion

Widespread vaccination against COVID-19 represents the current goal of public health. Allergists have been central insofar in order to define the minority of patients with contraindications to vaccination. Furthermore, as stated in the EAACI position paper, allergists should reassure patients with a severe allergic background in order to increase their compliance toward vaccines [5]. The so-called "infodemic", i.e., uncontrolled spreading of inflated news and fake news regarding COVID-19, has heightened anxiety concerning vaccine safety [18,19]. Anxiety has, in turn, long been regarded as an important issue in the allergic population due to the known long-lasting harmful effects on psychological balance observed in anaphylaxis survivors a [3].

Yet, the link between anxiety and allergy and their clinical consequences is complex and not entirely explored. Several researchers reported a pathophysiological link between anxiety and the onset of nocebo reactions. Higher anxiety states correlate whit nocebo during drug provocation tests [20,21]. Nocebos, in turn, can affect 78% of adverse reactions to anti-COVID-19 vaccines, according to a recent meta-analysis of randomized controlled trials [22].

Our study aimed to explore in detail the relationship between anxiety and allergy in the context of COVID-19 vaccination. Our data confirmed the correlation between a history of a severe allergic reaction, especially, to drugs, and anxiety. Specifically, we found that patients with severe allergies had a higher state anxiety score compared to patients with mild allergies. We also observed a non-significant trend towards higher levels of trait anxiety in patients with a severe allergy. The association between severe allergy and state anxiety was particularly strong in SDR patients.

On the other hand, the majority of subjects (especially those with a severe allergic background) felt reassured by an allergist-led protected setting. This highlights the importance of considering the psychological profile of allergic vaccinees and supports the role of allergists in overcoming vaccine hesitancy.

To our knowledge, the relationship between anxiety and allergy has not previously been evaluated in the context of COVID-19 vaccination. Nonetheless, our study does have limitations. First, our population was only constituted by allergic patients (both mild and severe) since, by definition, only this population was directed to our protected vaccinal sessions. Still, our aim was to describe the impact of anxiety in severely allergic patients, and therefore MAG represented a reasonable study comparator. In fact, significant differences were observed between these groups. Second, clinical and demographic data were self-reported and anonymized, preventing post hoc validation of acquired data or extension beyond predefined analyses. However, patient referral to vaccination in a protected setting was based on a physician review of individual clinical data. Third, while the STAY questionnaire is a validated and widely used tool in medical research, it does not cover the whole spectrum of anxiety and could underestimate the impact of some confounding factors. Moreover, we did not collect data about patient education level and knowledge or beliefs about vaccines or drugs. However, we detected no significant difference in state anxiety by vaccine dose number (i.e., primary cycle or booster doses). This could imply that even vaccine experience does not affect vaccine-related anxiety.

Fourth, the patient sample size was relatively small, and there was no long-term follow-up, limiting the detection of infrequent or delayed events and potential correlation with anxiety.

5. Conclusions

In summary, we showed how patients with a severe allergic background, especially severe drug allergy, have a significant psychological burden and concern about new vaccines. A protected setting led by an allergist not only could be effective in ensuring vaccination safety in patients with clinically relevant allergic history [8] but could also increase patient compliance toward vaccinations.

Supplementary Materials: The following supporting information can be downloaded at: https://www.mdpi.com/article/10.3390/vaccines10122047/s1, Figure S1: Answers to "Do you think that protected setting made you feel More Anxious, Less anxious or had no impact on your anxiety?" in the different subgroups; Figure S2: correlation between age range and state anxiety; Figure S3: correlation between age range and trait anxiety. Table S1: Demographic and clinical features.

Author Contributions: Conceptualization, M.-R.Y., G.A.R., V.E.D.M., S.Z. and M.M.; Formal Analysis, N.V., C.A., G.B., M.-R.Y. and G.A.R.; Investigation: B.B., C.A., G.B., S.N., M.B.C., M.R. and M.-R.Y.; Data Curation, N.V., G.A.R. and C.A.; Writing—Original Draft Preparation, G.B., C.A. and G.A.R.; Writing—Review & Editing, M.-R.Y., G.A.R., B.V., C.A. and G.B.; Supervision, L.D., G.A.R., F.B., V.E.D.M. and M.-R.Y. All authors have read and agreed to the published version of the manuscript.

Funding: This research received no external funding.

Institutional Review Board Statement: Ethical review and approval were waived for this study due to the anonymous nature of the study. Data collection was designed in compliance with the European Commission [Article 29 data protection working party (0829/14/EN WP216)] guidelines for anonymization in such a way that patient identification was impossible to the investigators or to other subjects.

Informed Consent Statement: Patient consent was waived due to the non-interventional and anonymous nature of the study. Data collection was designed in compliance with the European Commission [Article 29 data protection working party (0829/14/EN WP216)] guidelines for anonymization in such a way that patient identification was impossible to the investigators or to other subjects.

Data Availability Statement: The data presented in this study are available on request from the corresponding author. The data are not publicly available due to privacy.

Acknowledgments: All authors want to thank the staff of IRCCS San Raffaele Hospital's vaccination center.

Conflicts of Interest: The authors declare no conflict of interest.

References

1. Lee, J.K.; Vadas, P. Anaphylaxis: Mechanisms and management. *Clin. Exp. Allergy* **2011**, *41*, 923–938. [CrossRef] [PubMed]
2. Montoro, J.; Mullol, J.; Jáuregui, I.; Dávila, I.; Ferrer, M.; Bartra, J.; del Cuvillo, A.; Sastre, J.; Valero, A. Stress and Allergy. *J. Investig. Allergol. Clin. Immunol.* **2009**, *19* (Suppl. S1), 40–47. [CrossRef] [PubMed]
3. Baiardini, I.; Gaeta, F.; Molinengo, G.W.; Braido, F.; Canonica, G.; Romano, A. Quality-of-life issues in survivors to anaphylactic reactions to drugs. *Allergy* **2015**, *70*, 877–879. [CrossRef] [PubMed]
4. Altmann, D.M.; Boyton, R.J. COVID-19 vaccination: The road ahead. *Science* **2022**, *375*, 1127–1132. [CrossRef] [PubMed]
5. Sokolowska, M.; Eiwegger, T.; Ollert, M.; Torres, M.J.; Barber, D.; Del Giacco, S.; Jutel, M.; Nadeau, K.C.; Palomares, O.; Rabin, R.L.; et al. EAACI statement on the diagnosis, management and prevention of severe allergic reactions to COVID-19 vaccines. *Allergy* **2021**, *76*, 1629–1639. [CrossRef] [PubMed]
6. Shimabukuro, T.; Nair, N. Allergic Reactions Including Anaphylaxis After Receipt of the First Dose of Pfizer-BioNTech COVID-19 Vaccine. *JAMA-J. Am. Med. Assoc.* **2021**, *325*, 780. [CrossRef] [PubMed]
7. Klimek, L.; Jutel, M.; Akdis, C.A.; Bousquet, J.; Akdis, M.; Torres-Jaen, M.; Agache, I.; Canonica, G.W.; Del Giacco, S.; O'Mahony, L.; et al. ARIA-EAACI statement on severe allergic reactions to COVID-19 vaccines—An EAACI-ARIA Position Paper. *Allergy* **2020**, *76*, 1624–1628. [CrossRef] [PubMed]
8. Yacoub, M.-R.; Cucca, V.; Asperti, C.; Ramirez, G.A.; Della-Torre, E.; Moro, M.; Zandalasini, C.; Di Napoli, D.; Ambrosio, A.; Signorelli, C.; et al. Efficacy of a rational algorithm to assess allergy risk in patients receiving the BNT162b2 vaccine. *Vaccine* **2021**, *39*, 6464–6469. [CrossRef] [PubMed]

9. Chiang, V.; Saha, C.; Yim, J.; Au, E.Y.; Kan, A.K.; Hui, K.S.H.; Li, T.S.; Lo, W.L.W.; Hong, Y.D.; Ye, J.; et al. The Role of the Allergist in Coronavirus Disease 2019 Vaccine Allergy Safety: A Pilot Study on a "Hub-and-Spoke" Model for Population-Wide Allergy Service. *Ann. Allergy Asthma Immunol.* **2022**, *129*, 308–312.e1. [CrossRef] [PubMed]
10. Decreto Legge 12 Marzo 2021, Gazzetta Ufficiale n.72 del 24-03-2021. Available online: https://www.gazzettaufficiale.it/gazzetta/serie_generale/caricaDettaglio?dataPubblicazioneGazzetta=2021-03-24&numeroGazzetta=72 (accessed on 3 July 2022).
11. Filon, F.L.; Lazzarato, I.; Patriarca, E.; Iavernig, T.; Peratoner, A.; Perri, G.; Ponis, G.; Rocco, G.; Cegolon, L. Allergic Reactions to COVID-19 Vaccination in High-Risk Allergic Patients: The Experience of Trieste University Hospital (North-Eastern Italy). *Vaccines* **2022**, *10*, 1616. [CrossRef] [PubMed]
12. Banerji, A.; Wolfson, A.R.; Wickner, P.G.; Cogan, A.S.; McMahon, A.E.; Saff, R.; Robinson, L.B.; Phillips, E.; Blumenthal, K.G. COVID-19 Vaccination in Patients with Reported Allergic Reactions: Updated Evidence and Suggested Approach. *J. Allergy Clin. Immunol. Pract.* **2021**, *9*, 2135–2138. [CrossRef] [PubMed]
13. Turner, P.J.; Larson, H.; Dubé, È.; Fisher, A. Vaccine Hesitancy: Drivers and How the Allergy Community Can Help. *J. Allergy Clin. Immunol. Pract.* **2021**, *9*, 3568–3574. [CrossRef] [PubMed]
14. Bagos-Estevez, A.G.; Ledford, D.K. Anaphylaxis: Definition, Epidemiology, Diagnostic Challenges, Grading System. *Immunol. Allergy Clin. N. Am.* **2021**, *42*, 1–11. [CrossRef] [PubMed]
15. Article 29 Data Protection Working Party. 2014. Available online: https://ec.europa.eu/justice/article-29/documentation/opinion-recommendation/files/2014/wp216_en.pdf (accessed on 17 September 2022).
16. Julian, L.J. Measures of anxiety: State-Trait Anxiety Inventory (STAI), Beck Anxiety Inventory (BAI), and Hospital Anxiety and Depression Scale-Anxiety (HADS-A). *Arthritis Care Res.* **2011**, *63* (Suppl. S11), S467–S472. [CrossRef] [PubMed]
17. Knight, R.G.; Waal-Manning, H.J.; Spears, G.F. Some norms and reliability data for the State-Trait Anxiety Inventory and the Zung Self-Rating Depression scale. *Br. J. Clin. Psychol.* **1983**, *22 Pt 4*, 245–249. [CrossRef] [PubMed]
18. Vai, B.; Cazzetta, S.; Ghiglino, D.; Parenti, L.; Saibene, G.; Toti, M.; Verga, C.; Wykowska, A.; Benedetti, F. Risk Perception and Media in Shaping Protective Behaviors: Insights From the Early Phase of COVID-19 Italian Outbreak. *Front. Psychol.* **2020**, *11*, 563426. [CrossRef] [PubMed]
19. Bendau, A.; Plag, J.; Petzold, M.B.; Ströhle, A. COVID-19 vaccine hesitancy and related fears and anxiety. *Int. Immunopharmacol.* **2021**, *97*, 107724. [CrossRef] [PubMed]
20. Losappio, L.M.; Cappai, A.; Arcolaci, A.; Badiu, I.; Bonadonna, P.; Boni, E.; Bussolino, C.; Caminati, M.; Galati, P.; Heffler, E.; et al. Anxiety and Depression Effects During Drug Provocation Test. *J. Allergy Clin. Immunol. Pract.* **2018**, *6*, 1637–1641. [CrossRef] [PubMed]
21. Soyyiğit, Ş.; Aydın, Ö.; Yılmaz, I.; Özdemir, S.K.; Cankorur, V.; Atbaşoğlu, C.; Çelik, G.E. Evaluation of drug provocation test–related anxiety in patients with drug hypersensitivity. *Ann. Allergy Asthma Immunol.* **2016**, *117*, 280–284. [CrossRef] [PubMed]
22. Haas, J.W.; Bender, F.L.; Ballou, S.; Kelley, J.M.; Wilhelm, M.; Miller, F.G.; Rief, W.; Kaptchuk, T.J. Frequency of Adverse Events in the Placebo Arms of COVID-19 Vaccine Trials: A Systematic Review and Meta-Analysis. *JAMA Netw. Open* **2022**, *5*, e2143955. [CrossRef] [PubMed]

vaccines

MDPI

Systematic Review

Hearing Loss after COVID-19 and Non-COVID-19 Vaccination: A Systematic Review

Xin Wei Liew [1,†], Zer Han Malcolm Tang [1,†], Yan Qing Cherie Ong [1,†] and Kay Choong See [2,*]

1 Yong Loo Lin School of Medicine, National University of Singapore, Singapore 169856, Singapore; liewxinwei@gmail.com (X.W.L.); malcolmtangzh@gmail.com (Z.H.M.T.); cherieong379@gmail.com (Y.Q.C.O.)
2 Division of Respiratory and Critical Care Medicine, Department of Medicine, National University Hospital, Singapore 119228, Singapore
* Correspondence: kaychoongsee@nus.edu.sg
† These authors contributed equally to this work.

Abstract: (1) Background: Vaccine safety is an important topic with public health implications on a global scale. The purpose of this study was to systematically review available literature assessing sensorineural hearing loss (SNHL) incidence and severity following both coronavirus disease 2019 (COVID-19) and non-COVID-19 vaccinations, as well as prognosis and outcomes. (2) Methods: This systematic review was performed according to the Preferred Reporting Items for Systematic Review and Meta-Analysis guidelines. Relevant publications evaluating post-vaccination SNHL were selected from PubMed and Embase, searching from inception to July 2023. (3) Results: From 11 observational studies, the incidence of post-vaccination SNHL was low for both COVID-19 and non-COVID-19 vaccines, ranging from 0.6 to 60.77 per 100,000 person-years, comparable to all-cause SNHL. (4) Conclusions: The incidence rates of SNHL following COVID-19 and non-COVID-19 vaccinations remained reassuringly low. Most patients experienced improved hearing function in the weeks to months following vaccination. This study underscores the importance and safety of vaccinations and encourages ongoing surveillance and detailed reporting of hearing loss cases post-vaccination.

Keywords: vaccine; vaccination; hearing loss; hearing impairment; deaf; deafness

Citation: Liew, X.W.; Tang, Z.H.M.; Ong, Y.Q.C.; See, K.C. Hearing Loss after COVID-19 and Non-COVID-19 Vaccination: A Systematic Review. *Vaccines* **2023**, *11*, 1834. https://doi.org/10.3390/vaccines11121834

Academic Editor: Seth Pincus

Received: 17 October 2023
Revised: 7 December 2023
Accepted: 8 December 2023
Published: 9 December 2023

1. Introduction

Vaccination is one of the best public health interventions in modern times. Vaccines have successfully eradicated debilitating diseases such as smallpox and have also dramatically reduced the incidence rates of other major diseases such as polio and measles [1,2]. Annually, vaccinations are estimated to save 2–3 million lives [1]. With rapidly advancing scientific technologies, almost 30 microorganisms can be targeted with up to 70 vaccines and counting [3].

Since 1796, when Edward Jenner invented the first vaccine against smallpox, vaccinations have saved millions of lives and are indispensable in a physician's arsenal against microbiological diseases. Most recently, the vaccine has once again been relied upon, specifically for the novel coronavirus, severe acute respiratory syndrome coronavirus-2 (SARS-CoV-2). The World Health Organization (WHO) declared coronavirus disease 2019 (COVID-19) a pandemic on 11 March 2020, and great efforts were invested in producing an effective and safe COVID-19 vaccine [4]. Studies have shown that COVID-19 vaccination has also been pivotal in reducing the morbidity and mortality of COVID-19 patients [5].

Vaccine hesitancy is dangerous and not unique to COVID-19 vaccinations. Concerns about vaccine side effects are the second most common reason driving reluctance to take COVID-19 vaccinations [6,7]. Public concern about vaccine safety is expected and understandable. The WHO identified "reluctance or refusal to vaccinate despite the availability of vaccines" as one of the 10 threats to global health in 2019 [8]. Similarly, measles outbreaks in the United States (where endemic measles is eradicated) are largely contributed by intentional refusal to vaccinate [9]. It is estimated that a 5% decline in measles, mumps, and rubella (MMR) vaccine coverage in the United States would result in an estimated three-fold increase in measles cases for children aged 2 to 11 years annually [10]. This warning sign is found with pertussis, where vaccine hesitancy was linked to an increased risk for pertussis in some populations studied [9].

With the widespread uptake of COVID-19 vaccinations worldwide, otolaryngologic practices saw an increase in the number of anecdotal reports of sensorineural hearing loss (SNHL) post-vaccination [11–13]. Specialists in this field encounter increasing challenges in the counseling of such patients who report a temporal association of hearing loss post-COVID-19 vaccination, particularly in terms of the incidence, severity, and prognosis of the hearing loss.

SNHL is defined by the American Academy of Otolaryngology-Head and Neck Surgery as an acute 30 dB hearing loss across three consecutive frequencies as confirmed by audiometry [14], while hearing loss severity is graded based on pure tone audiogram hearing ranges (Table 1) [15]. In this paper, the definition of SNHL is expanded to include 26 dB hearing loss as per Clark et al.'s severity grading and SNHL diagnoses made by clinicians within the individual studies [15]. The annual incidence of SNHL was on average 27 per 100,000 person-years and ranges from 11 to 77 per 100,000 persons per year, depending on age [16]. There are various plausible etiologies for acquiring SNHL, including age-related, noise-related, drug-related, infection/inflammation, trauma, tumors, systemic disorders, vascular disorders, and vaccine-related [17].

Table 1. Severity of hearing loss based on audiogram metrics.

Hearing Range	dB
Normal	−10–25
Mild	26–40
Moderate	41–55
Moderately Severe	56–70
Severe	70–90
Profound	91+

dB: decibel.

The pathogenesis of how the COVID-19 vaccination causes hearing loss is not well understood. Proposed explanations include both the mRNA payload and the lipid nanoparticle delivery vehicle causing auto-immunogenicity [18] as well as the production of immunoglobulin G 10–14 days after vaccine administration, which coincided with SNHL 10–14 days after the vaccination.

A systematic review of the available current literature was therefore conducted to review the incidence and severity of sudden sensorineural hearing loss post-COVID-19 and non-COVID-19 vaccinations. We studied vaccines against hepatitis B, diphtheria, tetanus, measles, mumps, rubella, rabies, and influenza. We aimed to characterize this phenomenon further to guide clinical practice for all physicians, ranging from primary care physicians to otolaryngological clinical practitioners, so that physicians can provide public health messaging to minimize vaccine hesitancy. It is imperative to be up-to-date and transparent about the safety of vaccinations to best promote awareness and ultimately widespread acceptance of vaccines [19].

2. Materials and Methods

The study has been registered with PROSPERO (registration number: CRD42023441395). This systematic review was performed according to the Preferred Reporting Items for Systematic Review and Meta-Analysis (PRISMA) guidelines. A comprehensive search of PubMed (pubmed.ncbi.nlm.nih.gov, accessed on 30 July 2023) and Embase (embase.com, accessed on 30 July 2023) was conducted to identify the relevant literature (Figure 1).

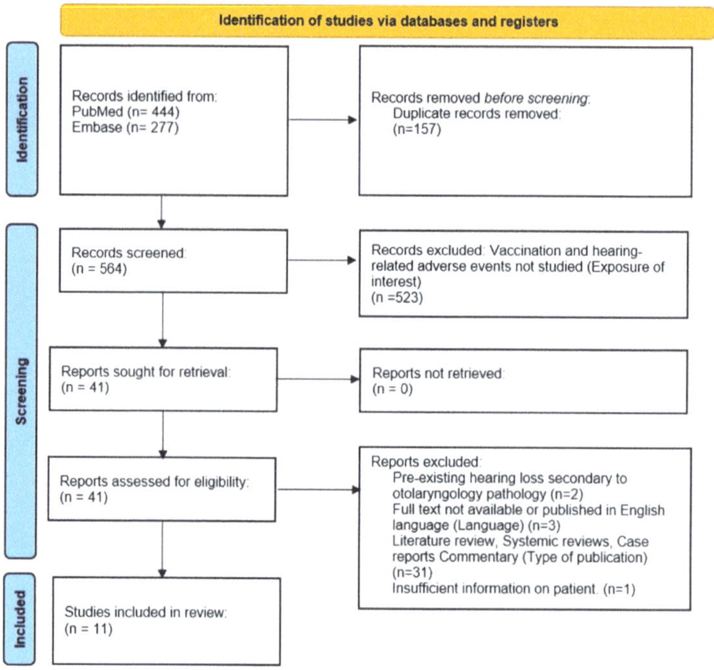

Figure 1. PRISMA flow diagram.

The keywords included "vaccine" and "vaccination" AND "hearing loss", "deaf", and "deafness". There was no limit to the timeframe of the search, which was performed on 30 July 2023.

The search produced a list of 561 unique articles. Screening of titles and abstracts was conducted, with analysis of the full texts if there were any doubts as to the suitability of the work for inclusion. We included all observational studies with a description of hearing loss, including those where quantitative audiogram measurements were not listed. A key exclusion criterion included the study population, which already had pre-existing otologic disorders affecting baseline hearing. We have filtered the number of papers to 11.

A qualitative review of the included studies was then performed to uncover a general understanding of the associations of vaccine exposure with hearing loss, as well as the incidence and severity. In addition, interventions to manage hearing loss after vaccination as well as patients' outcomes were studied.

In our carefully selected observational studies, we reviewed the incidence or prevalence of hearing loss in patients receiving COVID-19 and non-COVID-19 vaccinations. Additional data fields extracted from the full-text documents included the following: patient demographic, vaccine type, number of patients who received the vaccine, time of onset of SNHL since vaccination, associated symptoms, and treatment initiated. XWL, YQCO, and ZHMT did the full text screen and data extraction. XWL and ZHMT assessed observational studies for bias using ROBINS-I (Table 2). Any discrepancies were solved through discussion with the senior author, KCS.

Table 2. Risk of bias assessment of observational studies using ROBINS-I.

Study Author (Year of Publication) (Ref.)	Vaccine Type	Bias Due to Confounding	Bias in Selection of Participants	Bias in Classification	Bias Due to Deviations	Bias Due to Missing	Bias in the Measurement of Outcomes	Bias in Reporting of Data	Overall Risk of Bias
Astrayan (2008) [20]	MMR	Moderate	Moderate	Low	NI	Moderate	Moderate	Low	Moderate
Baxter (2016) [21]	Influenza, tetanus, reduced diphtheria, reduced acellular pertussis, and zoster	Moderate	Low	Low	NI	Moderate	Moderate	Low	Moderate
Avci (2021) [22]	COVID-19	Moderate	Low	Low	Low	Low	Low	Low	Low
Filippatos 2021 [23]	COVID-19	Moderate	Low	Low	NI	Moderate	Moderate	Low	Moderate
Wichova (2021) [11]	COVID-19	Moderate	Low	Low	Low	Moderate	Low	Low	Low
Chen (2022) [24]	COVID-19	Moderate	Low	Low	Low	Low	Low	Low	Low
Formeister (2022) [25]	COVID-19	Moderate	Low	Low	NI	Moderate	Moderate	Low	Moderate
Guo (2022) [26]	COVID-19	Moderate	Low	Low	NI	Moderate	Low	Low	Moderate
Yanir (2022) [27]	COVID-19	Moderate	Low	Low	NI	Moderate	Low	Low	Moderate
Nieminen (2023) [28]	COVID-19	Moderate	Low	Low	NI	Moderate	Low	Low	Moderate
Thai-Van (2023) [29]	COVID-19	Moderate	Low	Low	NI	Moderate	Low	Low	Moderate

NI = no information.

3. Results

Out of 444 studies extracted from PubMed and 277 from Embase, we have included 11 observational studies. The PRISMA flowchart is displayed in Figure 1, and the quality evaluation results are displayed in Table 2.

From Table 2, bias across observational studies is mostly "moderate" overall. This is largely due to reporting bias from electronic records, with minimal to no effort in reducing confounders in analysis. Larger observational studies utilize self-reporting systems, as seen with the Vaccine Adverse Event Reporting System (VAERS) in the United States [20,24–26] and national healthcare registries in Finland [28], Israel [27], and France [29]. Therefore, reporting bias remains a problem in interpreting the results. Furthermore, confounders play a large role in data analysis, especially when various other demographic and medical factors have to have a direct impact on SNHL, such as cardiovascular risk factors [30]. While the bias has been evaluated as "moderate", we continued to include these studies due to the fulfillment of our inclusion criteria after filtering from database searches, and we believe they contribute to the available body of evidence pertaining to the limited study of post-vaccination SNHL.

3.1. Observational Studies of COVID-19 Vaccines

A total of nine observational studies focused on COVID-19 vaccines (Table 3). The COVID-19 vaccines available use mRNA (e.g., manufactured by Pfizer or Moderna), viral vector-based (e.g., manufactured by Johnson and Johnson), or inactivated vaccine platforms (e.g., manufactured by Sinovac or AstraZeneca). A majority of studies evaluated mRNA-based vaccines only [11,27,29], while others also included viral vector-based vaccines [24–26] and inactivated virus vaccines [22,28] in their studies.

Generally, all nine studies showed that the incidence and prevalence of SNHL associated with COVID-19 vaccination were very low, even across different demographics and vaccine types. A range of incidence in large-scale studies can be appreciated, from as low as 0.6 to 28.0 cases per 100,000 person-years. For Yanir et al., however, a study carried out on the Israeli population suggested an increasing trend of SNHL post-vaccination as compared to previous years prior to the vaccination [27]. The paper found an increasing incidence rate (IR) of 60.77 (95% CI, 48.29–73.26) per 100,000 person-years post-COVID-19 vaccination as compared to previous reference years prior to COVID-19 vaccination, which demonstrated an IR of 41.50 (95% CI, 37.98–45.01) per 100,000 person-years in 2018 and 44.46 (95% CI, 40.85–48.07) per 100,000 person-years in 2019.

The range of prevalence appears to be wide, ranging from as low as 0.00324% in Yanir et al.'s study to 3.85% in Wichova et al.'s study [11,27]. While smaller-scale observational studies conducted by Wichova, Filippatos, and Avci [11,22,23] demonstrated a higher prevalence of SNHL, beginning at 0.2% in Filippatos et al.'s study [23], the larger observational studies from the USA, Finland, France, and Israel show that the nationwide prevalence of SNHL is reassuringly low, with the highest prevalence being 0.0142% in Nieminen's study [28].

Most studies did not specify the severity of SNHL or discuss more about the recovery and prognosis of those who did suffer from SNHL. Uniquely, hearing loss after COVID-19 vaccination was seen in 1.2% of patients with COVID-19 infection in the past 6 months, as compared to only 0.1% of patients without COVID-19 infection [22].

Table 3. Observational studies of SNHL in patients after taking COVID-19 vaccines.

Vaccine Type	Study Author (Year of Publication)	Number of Participants/Doses	Mean Age at the Time of Vaccination	Incidence of SNHL	Prevalence of SNHL	Severity of SNHL	Prognosis
COVID-19 (mRNA and viral vector) (Pfizer-BioNTech, Moderna, or Janssen/Johnson and Johnson)	Chen (2022) [24]	224,660,453 participants from the Vaccine Adverse Event Reporting System (VAERS). Hearing impairment is defined as SNHL, aural fullness, and tinnitus.	NI	6.66 per 100,000 person-years (14,956 reports over a ~2-year period of follow-up)	0.00666% (14,956 reports)	NI	NI
COVID-19 (mRNA and viral vector) (Pfizer-BioNTech, Moderna, or Janssen/Johnson and Johnson)	Formeister (2022) [25]	185,424,899 doses from the Vaccine Adverse Event Reporting System (VAERS)	54 years	0.6 to 28.0 per 100,000 person-years (555 reports over a 7-month follow-up period)	0.16 cases per 100,000 doses of both the Pfizer-BioNTech and Moderna vaccines 0.22 cases per 100,000 doses of Janssen/Johnson and Johnson vaccines	NI	8 of 14 patients with posttreatment audiometric data experienced improvement after receiving treatment
COVID-19 (mRNA) (Pfizer-BioNTech)	Yanir (2022) [27]	Israel First dose 2,602,557 participants	46 years	60.77 per 100,000 person-years (91 reports over a 6 month follow-up period)	0.00350% (91 reports)	NI	NI
		Second dose 2,441,719 participants		56.24 per 100,000 person-years (79 reports over a 6 month follow-up period)	0.00324% (79 reports)		

Table 3. *Cont.*

Vaccine Type	Study Author (Year of Publication)	Number of Participants/Doses	Mean Age at the Time of Vaccination	Incidence of SNHL	Prevalence of SNHL	Severity of SNHL	Prognosis
COVID-19 (mRNA) (Pfizer-BioNTech of Moderna)	Thai-Van (2023) [29]	France 97,840,529 doses of Pfizer 22,690,889 doses of Moderna from the Natural Healthcare Registry	51 years (for Pfizer) 47 years (for Moderna)	Pfizer	Pfizer	NI	NI
				1.45 per 1,000,000 injections	0.000145% (142 cases in 97,840,529 doses)		
				Moderna	Moderna		
				1.67 per 1,000,000 injections	0.000128% (29 cases in 22,690,889 doses)		
COVID-19 (mRNA and inactivated) (Pfizer-BioNTech, Moderna, or AztraZeneca)	Nieminen (2023) [28]	Finland ~5,500,000 individuals from the respective national registry	NI	AstraZeneca	AstraZeneca	NI	NI
				22.1 per 100,000 person-years (71 reports over ~2-year period of follow-up)	0.00130% (71 reports)		
				Pfizer	Pfizer		
				21.2 per 100,000 person-years (779 reports over ~2-year period of follow-up)	0.0142% (779 reports)		
				Moderna	Moderna		
				18.5 per 100,000 (188 reports over ~2-year period of follow-up person-years	0.00452% (188 reports)		

Table 3. *Cont.*

Vaccine Type	Study Author (Year of Publication)	Number of Participants/Doses	Mean Age at the Time of Vaccination	Incidence of SNHL	Prevalence of SNHL	Severity of SNHL	Prognosis
COVID-19 (inactivated) (CoronaVac or Sinovac Life Sciences)	Avci (2021) [22]	1710 participants (healthcare workers)	35.79 years	NI	0.3% (5 reports)	NI	NI
COVID-19 (mRNA) (Pfizer)	Filippatos (2021) [23]	502 participants	48.17 years	NI	0.2% (1 report)	NI	Improvement noted
COVID-19 (mRNA) (Pfizer-BioNTech or Moderna)	Wichova (2021) [11]	1641 clinical visits (in 2020)	60.9 years	NI	2020 2.44% (40 reports)	The mean pure tone average (PTA) was 52.2 ± 30.6 dB HL for the affected ear and 21.2 ± 12.5 dB HL for the unaffected ear. The word recognition score (WRS) was 60.6 ± 38% for the affected ear and 90 ± 23.0% for the unaffected ear.	NI
		1325 clinical visits (in 2021)			2021 3.85% (51 reports)		
COVID-19 (mRNA and viral vector) (Pfizer-BioNTech, Moderna, or Janssen/Johnson and Johnson)	Guo (2022) [26]	717,577 adverse effects from the Vaccine Adverse Event Reporting System (VAERS)	NI	NI Of the 717,577 vaccine adverse effects: Deafness: 809 incident reports, PRR 2.03 Hypoacusis: 781 incident reports, PRR 2.50	NI	NI	NI

NI = no information.

3.2. Observational Studies on Non-COVID-19 Vaccines

Only two large-scale observational studies studied non-COVID-19 vaccines and SNHL. Asatryan et al. studied the measles, mumps, and rubella (MMR) vaccine, and Baxter et al. studied the influenza, tetanus, reduced diphtheria, reduced acellular pertussis, and zoster vaccines [20,21].

In Asatryan et al., the incidence of hearing loss reported after vaccination (1 per 6–8 million doses) appears to be substantially rarer than that seen after natural measles or mumps infection (1 per 20,000 infections) [20]. For Baxter et al., across 7 years of follow-up and over 23 million vaccines, patients with the development of SNHL were not associated with immunization [21]. The severity of SNHL was not described in these two studies.

4. Discussion

4.1. Overview of Results

This comprehensive systematic review of COVID-19 and non-COVID-19 vaccinations and SNHL aims to better elucidate the complications of administering such vaccines. With ongoing controversy about the effectiveness and safety of vaccines, especially amongst population groups advocating against vaccinations, it is the duty of the medical and scientific community to keep everyone informed on the most accurate and up-to-date data on vaccine safety.

In this systematic review, we found that both the incidence and prevalence of SNHL after COVID-19 vaccinations were low, corresponding to a low disease burden and pressure. The incidence range of SNHL is low, ranging from 0.6 to 60.77 per 100,000 person-years across various cohort studies, and very few case reports on SNHL exist relative to the large number of vaccines administered. In large-scale observational studies, the incidence from all papers reviewed demonstrated that the incidence of SNHL was mostly compatible with the average annual incidence of 27 per 100,000 person-years for all causes, as reported by Alexander et al. in 2013, studying 60 million patients from the United States (US) across 2006–2007 [16]. In nationwide studies conducted in the US, Formeister et al. reported an incidence ranging from 0.6 to 28.0 cases per 100,000 person-years, whereas Chen reported an incidence of 6.66 per 100,000 person-years [24,25]. Similarly, the Finnish study by Nieminen et al. shows an incidence of 21.2 to 22.1 cases per 100,000 person-years, depending on which vaccine was used [28]. Thai-Van et al.'s study similarly demonstrated a small incidence of 1.45 or 1.67 reports per 1,000,000 vaccinations [29]. These large-scale observational studies are in keeping with the average annual incidence of 27 per 100,000 person-years from all causes.

However, there exists an exception in Yanir's study on the Israeli population in 2022 where the incidence ratio (IR) of SNHL after COVID-19 vaccines was a high of 60.77 (95% CI, 48.29–73.26) per 100,000 person-years, averaged across age groups, and this was comparatively higher than the other observational studies [27]. Across age, the IR increased from 22.44 to 150.53 per 100,000 person-years from age groups 16–44 to patients older than 65. Similarly, Alexander's study also revealed an increasing incidence of SNHL with age, from 11 per 100,000 for patients younger than 18 years to 77 per 100,000 for patients 65 years and older, and established a positive correlation between age and the incidence of sensorineural hearing loss, which is in keeping with Yanir's study [16,27]. Additionally, Yanir's relatively high IR of SNHL is also found in previous reference years prior to the COVID-19 pandemic in the same paper: 41.50 (95% CI, 37.98–45.01) per 100,000 person-years in 2018 and 44.46 (95% CI, 40.85–48.07) per 100,000 person-years in 2019, suggesting that the baseline incidence of SNHL in the Israeli population is already above the average proposed by Alexander [16,27]. Yanir noted in his study that people who were vaccinated were older and may be sicker than the reference population, with a mean age of 46.8 ± 19.6 years [27]. Many reasons account for this higher IR of post-vaccination SNHL in Yanir's cohort relative to similar studies [27]. These include the inherent differences between the Israel population and other study populations and many health confounders that were not accounted for. In particular, Yanir noted that cardiovascular risk factors as well as coagulation disorders, which are themselves risk factors for SNHL, were not accounted for in the study and stated that the lack of data detailing the health characteristics of the exposure group was a serious limitation [27]. Additionally, Alexander's incidence, used as a reference in this paper, was calculated with patients from 2006–2007, and changes in health-seeking behavior over time could be attributed to the stark difference [16]. Fortunately, Yanir's study concludes with a small attributable risk (AR) to post-vaccination SNHL, with the highest AR of 3.74 per 100,000 vaccinees, and concludes that the influence on public health would be relatively minor [27].

In smaller observational studies ranging from 500 to 1710 participants, the incidence of SNHL was not reported, and there was no period of follow-up in those studies. Hence, these studies were not taken into account when reviewing the incidence.

All observational studies on COVID-19 vaccines showed that the incidence of SNHL associated with COVID-19 vaccination is reassuringly low, even across different demographics and vaccine types (Table 3). Similarly, the two studies on non-COVID-19 large-scale vaccination campaigns, such as the live attenuated MMR vaccine, the inactivated influenza vaccine, tetanus, reduced diphtheria, and reduced acellular pertussis (Tdap) vaccine, as well as the zoster vaccine, also did not demonstrate an increased incidence of SNHL in the general population (Table 4) [20,21].

The prevalence of SNHL after the COVID-19 vaccination appears reassuringly low as well. For the large-scale observational studies, the three studies comparing SNHL to the number of participants affected reflect a small range of prevalence from 0.00350% to 0.0142%. Nieminen's Finnish study reflected the highest prevalence, where 0.0142% of participants were found to have SNHL after the Pfizer-BioNTech COVID-19 vaccination [28]. Chen's study follows next, with a prevalence of 0.00666% [24]. Lastly, Yanir's Israeli study had the lowest prevalence of 0.00350% after the first dose of the COVID-19 vaccination [27].

For the other two large-scale observational studies comparing SNHL to the number of doses, the prevalence of SNHL was low as well. In Formeister's study, the prevalence was 0.16 cases per 100,000 doses for both the Pfizer-BioNTech and Moderna vaccines and 0.22 cases per 100,000 doses for the Janssen/Johnson and Johnson vaccine [25]. In Thai-Van's French study, the prevalence ranges from 0.000128% to 0.000145%, depending on the vaccine used [29].

In Guo's study comparing the adverse effects of the COVID-19 vaccination, the prevalence of deafness and hypoacusis only accounted for a small percentage, with 809 incident reports of deafness and 781 reports of hypoacusis out of 717,577 reported vaccination adverse effects [26]. Both account for only 0.1% of all vaccination adverse effects. Hence, even the prevalence of hearing loss within the pool of reported adverse effects post-COVID-19 vaccination is extremely low.

The prevalence of SNHL after COVID-19 vaccination in smaller observational studies appears to be significantly higher. In Filippatos' study of 502 healthcare workers, there was one case of SNHL, leading to a prevalence of 0.2% [23]. In Avci's study of 1710 healthcare workers, there were five reports of SNHL, giving a prevalence of 0.3%. Lastly, in Wichova's study, 40 of the 1641 patients and 51 of the 1325 patients who visited the clinic, respectively, in 2020 and 2021, were found to have clinically diagnosed SNHL, giving a prevalence of 2.44% (in 2020) and 3.85% (in 2021) [11,22]. While these prevalences appear to be higher than those reported in the large-scale observational studies, this can be largely accounted for due to a small sample size, which does not accurately represent the entire population, as well as selection bias. Avci's and Filippatos' studies were carried out exclusively on healthcare workers, which is epidemiologically not representative of the general population [22,23]. For Wichova's study, it too follows that there would be a proportionally greater number of individuals presenting at otolaryngologic clinics or participating in interviews who have hearing loss as compared to the general population, creating a selection bias [11]. The results may also be confounded by predisposing otolaryngologic pathologies, which may explain the hearing loss.

Table 4. Observational studies of SNHL in patients after taking non-COVID-19 vaccines.

Vaccine Type	Study Author (Year of Publication)	Number of Participants/Doses	Mean Age at the Time of Vaccination	Description of Incidence, Incidence Ratio (IR), Odds Ratio (OR) of SNHL	Prevalence Rate Ratio (PRR) of SNHL	Severity of SNHL	Prognosis
MMR	Asatryan (2008) [20]	168–224 million doses of MMR vaccine from 1990 to 2003	16 months	NI	59 cases—14 from VAERS and 15 from case reports; C1 case per 6–8 million doses	NI	NI
Influenza, tetanus, reduced diphtheria, reduced acellular pertussis, and zoster	Baxter (2016) [21]	8,354,237 doses given from 2007 to 2013	NI	ORs for vaccination 1 week prior to SSHL were: 0.965 (95% CI, 0.61–1.50) for trivalent inactivated influenza vaccine (TIV), 0.842 (95% CI, 0.39–1.62) for tetanus, reduced diphtheria, and reduced acellular pertussis, and 0.454 (95% CI, 0.08–1.53) for the zoster vaccine.	NI	NI	NI

NI = no information.

From the above, there does not appear to be a correlation between SNHL and COVID-19 vaccination, with the incidence as well as prevalence of SNHL post-vaccination being low across many large observational studies on different geographical populations. As such, the burden of SNHL post-vaccination, if any, is low and will likely remain low in time to come.

While incidence remained low across time and between countries, an observational study by Wichova et al. noted a pattern of increase in incidence since the pandemic [11]. Following the pandemic onset in early 2020 to the present, there has been a clear increase in this diagnosis, with a more than two-fold increase to 2.44 and 3.85% in 2020 and 2021, respectively. While an increased incidence does not by itself prove causation, the trend here does bring up concern that in some patients, there may be a post-vaccination change in hearing. One study compared the incidence of SNHL between the different vaccines [24]. Chen et al. identified increased risk for hearing disorder following administration of COVID-19 vaccines (both mRNA and virus vector) compared to influenza vaccination in real-world settings [24]. While incidences within each vaccine remain low and insignificant, the inter-vaccine differences could hold immunologic and biological mechanisms to uncover.

Our systematic review also explored the frequency and distribution of age in SNHL post-vaccination by comparing the mean age as well as the age range of the study populations at the time of vaccination. One intriguing aspect of this discussion is the trend observed in some studies, suggesting a rise in the age of individuals experiencing SNHL post-COVID-19 vaccination. Publications reported that the mean age at the time of vaccination is most prevalent in individuals greater than 45 years old [11,23,25,27,29,31,32]. This is consistent with the health patterns of the general population, in which SNHL prevalence increases with age. It is essential to consider age-related hearing loss as a baseline, as older adults may experience hearing loss coincidentally with receiving the vaccine, making it challenging to establish a direct causation. This trend raises critical questions about the interplay between age, vaccination, and hearing health.

There are various case reports and series on patients suffering from SNHL post-COVID-19 vaccination. Most reported mild to moderate hearing loss with complete recovery after corticosteroids, but there remained a handful reporting more severe SNHL, which had only a partial response to treatment. However, these case reports and series were excluded from our analysis, as these publications are primarily descriptive with no long-term follow-up data. The level of evidence is low, and it is likely that these reports represent a biased subgroup, where there is a risk of reporting bias. Our comprehensive study of observational studies, on the other hand, offers a broader perspective by analyzing patterns and trends across a larger population. This observational data provides a statistical basis to draw conclusions about the prevalence and incidence of SNHL post-vaccinations spanning across diverse demographics and may give a better perspective on the issue of SNHL in light of these case reports. Nonetheless, case reports are invaluable for elucidating rare conditions and atypical presentations of hearing deficits post-vaccinations. Kahn et al. reported on an alarming case of a young, 20-year-old male with bilateral profound SNHL as part of a multisystem inflammation and organ dysfunction of unknown mechanism after administration of the Pfizer COVID-19 vaccine [31]. In this patient, acute stroke, pericardial effusion and tamponade, pleural effusion, and acute kidney injury were described. This detailed narrative provides a foundation for further investigation and sheds light on the devastating, albeit rare, complications that can arise post-vaccination. Regardless, case reports suggest that the prognosis for post-vaccination SNHL was generally favorable, with frequent reversibility and partial to complete recovery in most cases.

4.2. Limitations of Our Study and Literature

Our systematic review has several limitations that should be acknowledged. For each vaccine, the number of studies available is limited, with few observational studies. Nevertheless, there is internal consistency in the overall conclusion of the papers included, and no prospective studies or randomized controlled trials exist for inclusion to dissuade our conclusion.

Another limitation is the inadequate data presented in the included studies. On top of the inherent bias of such articles evaluated as "moderate", these large-scale studies do not describe the severity, duration, and prognosis of SNHL post-vaccination. Importantly, the time of onset from SNHL. Additionally, another limitation is that, as the above studies are retrospective in nature, we cannot definitively conclude the causation between the vaccination and SNHL. Hence, we would recommend large-scale randomized control trials to support our theory and establish a more concrete understanding of the adverse effects of vaccines. Future studies and trials of vaccine safety should specifically check for this complication through objective hearing screening, with confirmation and data registration with pure tone audiometry in symptomatic patients.

A special consideration highlighted is the use of COVID-19 vaccines in a patient who previously contracted the COVID-19 virus itself. Avci et al. showed that the incidence of otolaryngology-specific symptoms such as hearing loss may be higher after inactivated COVID-19 vaccination in patients who were already previously infected by COVID-19 [22]. It may be prudent to inform such patients of a potentially increased risk relative to the general population before receiving an inactivated COVID-19 vaccine. The papers studied in our systematic review did not stratify the population into patients with prior COVID-19 infection compared to those without, and this may be a significant confounder for further researchers to elucidate its influence, even by non-COVID-19 infective agents.

Fortunately, there is nothing to suggest a direct association between SNHL and the administering of vaccines themselves, and vaccination campaigns with strong uptake in vaccination in the name of public health should continue to be encouraged. While there could be a potential link between vaccinations and SNHL, the evidence is largely anecdotal, and no correlation has been proven so far. Therefore, the benefit of mass vaccination in the aftermath of the SARS-CoV-2 pandemic remains unchallenged. As physicians responsible for the long-term health of our patients, information and the prognosis of the uncommon SNHL are important for us to aid our patients. While the incidence is fortunately low, future studies and reports on such complications should include detailed data on the illness for us to minimize the debilitating effects that deafness could potentially have. Another nuance to consider is that the available data primarily represents early post-vaccination periods, and the long-term effects of vaccinations (especially novel ones like COVID-19 vaccinations) on hearing loss are yet to be fully understood. Additionally, the heterogeneity of vaccine types, dosages, and booster scheduling across the included reports further complicates the interpretation of the findings.

4.3. Future Directions

Vaccine development, approval, and public acceptance are often a long process in which up-to-date and comprehensive data on potential complications and adverse effects is paramount to protecting society from diseases. We are relieved to have found low incidences of post-vaccination SNHL. Subsequently, various suggestions are raised to better elucidate the unknown variables. This can include more detailed logging of the severity of SNHL as well as its recovery and prognosis. More factors (largely under-recorded) with utility include the confounding effect of previously infected patients and various permutations of vaccination status (in terms of the number of booster shots received and duration between vaccinations). We theorize that deeper analysis of such factors can uncover unknown associations with adverse effects of vaccinations to better direct vaccine indications and even scheduling. We look forward to large-scale prospective randomized controlled trials with meaningful stratification of age, sex, and medical co-morbidities to conclude a causative effect more strongly between vaccinations and SNHL.

5. Conclusions

In conclusion, this review of 11 observational studies demonstrated a minimal co-relationship between vaccinations and SNHL. The incidence, prevalence, and hence burden of SNHL post-COVID-19 vaccinations remain small across many different nations. The majority of the observational studies report an incidence that falls within the average annual incidence of SNHL of 27 per 100,000 person-years for all causes in a large US study [16]. The prevalence of SNHL remains reassuringly low with the exception of small-scale observational studies, which can be accounted for by sampling bias and selection bias. Hence, the burden on SNHL is small and is likely to remain small with time.

Unilateral hearing loss seems to be more common than bilateral, and alert physicians can rely on the speedy usage of steroids as a safe and reliable treatment for SNHL to likely ameliorate patients' hearing impairment post-vaccination. Thankfully, the majority of patients will return to their normal level of hearing within weeks or months. Vaccinations and their protection for the global community strongly outweigh the weakly related otologic complications. International collaboration between otolaryngologists, immunologists, and vaccine researchers would further strengthen our knowledge in the area of post-vaccination hearing loss.

Author Contributions: Conceptualization, X.W.L., Y.Q.C.O., Z.H.M.T. and K.C.S.; data curation, X.W.L., Y.Q.C.O. and Z.H.M.T.; investigation, X.W.L. and Y.Q.C.O.; methodology, X.W.L., Y.Q.C.O., Z.H.M.T. and K.C.S.; supervision, K.C.S.; visualization, X.W.L.; writing—original draft, X.W.L., Y.Q.C.O. and Z.H.M.T.; writing—review and editing, X.W.L., Y.Q.C.O., Z.H.M.T. and K.C.S. All authors have read and agreed to the published version of the manuscript.

Funding: This research received no external funding.

Institutional Review Board Statement: Not applicable.

Informed Consent Statement: Not applicable.

Data Availability Statement: Data supporting this review are available in the reference section.

Conflicts of Interest: The authors declare no conflict of interest.

References

1. Delany, I.; Rappuoli, R.; De Gregorio, E. Vaccines for the 21st century. *EMBO Mol. Med.* **2014**, *6*, 708–720. [CrossRef] [PubMed]
2. Tognotti, E. The eradication of smallpox, a success story for modern medicine and public health: What lessons for the future? *J. Infect. Dev. Ctries.* **2010**, *4*, 264–266. [CrossRef] [PubMed]
3. Landry, S.; Heilman, C. Future directions in vaccines: The payoffs of basic research. *Health Aff.* **2005**, *24*, 758–769. [CrossRef] [PubMed]
4. WHO. Considerations for the Assessment of COVID-19 Vaccines for Listing by WHO. 2020. Available online: https://www.who.int/publications/m/item/considerations-for-the-assessment-of-covid-19-vaccines-for-listing-by-who (accessed on 26 November 2022).
5. Liu, Q.; Qin, C.; Liu, M.; Liu, J. Effectiveness and safety of SARS-CoV-2 vaccine in real-world studies: A systematic review and meta-analysis. *Infect. Dis. Poverty* **2021**, *10*, 132. [CrossRef] [PubMed]
6. Khairat, S.; Zou, B.; Adler-Milstein, J. Factors and reasons associated with low COVID-19 vaccine uptake among highly hesitant communities in the US. *Am. J. Infect. Control* **2022**, *50*, 262–267. [CrossRef] [PubMed]
7. Pires, C. Global Predictors of COVID-19 Vaccine Hesitancy: A Systematic Review. *Vaccines* **2022**, *10*, 1349. [CrossRef] [PubMed]
8. WHO. Ten Threats to Global Health in 2019. 2019. Available online: https://www.who.int/news-room/spotlight/ten-threats-to-global-health-in-2019#:~:text=This%20watchlist%20for%20priority%20research,that%20could%20cause%20a%20serious (accessed on 26 November 2022).
9. Phadke, V.K.; Bednarczyk, R.A.; Salmon, D.A.; Omer, S.B. Association Between Vaccine Refusal and Vaccine-Preventable Diseases in the United States: A Review of Measles and Pertussis. *JAMA* **2016**, *315*, 1149–1158. [CrossRef]
10. Lo, N.C.; Hotez, P.J. Public Health and Economic Consequences of Vaccine Hesitancy for Measles in the United States. *JAMA Pediatr.* **2017**, *171*, 887–892. [CrossRef]
11. Wichova, H.; Miller, M.E.; Derebery, M.J. Otologic Manifestations After COVID-19 Vaccination: The House Ear Clinic Experience. *Otol. Neurotol.* **2021**, *42*, e1213–e1218. [CrossRef]
12. Medina, M.C.; Gómez, M.R. Tinnitus, Sudden Sensorineural Hearing Loss, and Vestibular Neuritis As Complications of the Astra Zeneca COVID-19 Vaccine. *Cureus* **2022**, *14*, e20906.

13. Jeong, J.; Choi, H.S. Sudden sensorineural hearing loss after COVID-19 vaccination. *Int. J. Infect. Dis.* **2021**, *113*, 341–343. [CrossRef] [PubMed]
14. Chandrasekhar, S.S.; Tsai Do, B.S.; Schwartz, S.R.; Bontempo, L.J.; Faucett, E.A.; Finestone, S.A.; Hollingsworth, D.B.; Kelley, D.M.; Kmucha, S.T.; Satterfield, L.; et al. Clinical Practice Guideline: Sudden Hearing Loss (Update). *Otolaryngol. Head Neck Surg.* **2019**, *161* (Suppl. S1), S1–S45. [CrossRef] [PubMed]
15. Clark, J.G. Uses and abuses of hearing loss classification. *ASHA* **1981**, *23*, 493–500. [PubMed]
16. Alexander, T.H.; Harris, J.P. Incidence of sudden sensorineural hearing loss. *Otol. Neurotol.* **2013**, *34*, 1586–1589. [CrossRef] [PubMed]
17. Li, S.; Zhang, F.; Yang, J. Interpretation of AAOdouble ended arrowHNSF clinical practice guideline: Sudden hearing loss(update) and comparison with previous edition. *J. Clin. Otorhinolaryngol. Head Neck Surg.* **2020**, *34*, 385–388. [PubMed]
18. Tsilingiris, D.; Vallianou, N.G.; Karampela, I.; Liu, J.; Dalamaga, M. Potential implications of lipid nanoparticles in the pathogenesis of myocarditis associated with the use of mRNA vaccines against SARS-CoV-2. *Metab. Open* **2022**, *13*, 100159. [CrossRef] [PubMed]
19. Dube, E.; Ward, J.K.; Verger, P.; MacDonald, N.E. Vaccine Hesitancy, Acceptance, and Anti-Vaccination: Trends and Future Prospects for Public Health. *Annu. Rev. Public Health* **2021**, *42*, 175–191. [CrossRef] [PubMed]
20. Asatryan, A.; Pool, V.; Chen, R.T.; Kohl, K.S.; Davis, R.L.; Iskander, J.K. Live attenuated measles and mumps viral strain-containing vaccines and hearing loss: Vaccine Adverse Event Reporting System (VAERS), United States, 1990–2003. *Vaccine* **2008**, *26*, 1166–1172. [CrossRef]
21. Baxter, R.; Lewis, N.; Bohrer, P.; Harrington, T.; Aukes, L.; Klein, N.P. Sudden-Onset Sensorineural Hearing Loss after Immunization: A Case-Centered Analysis. *Otolaryngol. Head Neck Surg.* **2016**, *155*, 81–86. [CrossRef]
22. Avcı, H.; Karabulut, B.; Eken, H.D.; Faraşoğlu, A.; Çakil, T.; Çoruk, S.; Özel, H.; Kaya, N.K.; Özbalta, S. Otolaryngology-Specific Symptoms May Be Highly Observed in Patients With a History of Covid-19 Infection After Inactivated Coronavirus Vaccination. *Ear Nose Throat J.* **2021**, *102*, 1455613211028493. [CrossRef]
23. Filippatos, F.; Tatsi, E.-B.; Dellis, C.; Dessypris, N.; Syriopoulou, V.; Michos, A. Association of clinical and epidemiological characteristics with COVID-19 BNT162b2 mRNA vaccine short-term adverse reactions in healthcare workers. *Hum. Vaccines Immunother.* **2021**, *17*, 4755–4760. [CrossRef] [PubMed]
24. Chen, C.; Fu, F.; Ding, L.; Xiao, J. Hearing disorder following COVID-19 vaccination: A pharmacovigilance analysis using the Vaccine Adverse Event Reporting System. *J. Clin. Pharm. Ther.* **2022**, *47*, 1789–1795. [CrossRef] [PubMed]
25. Formeister, E.J.; Wu, M.J.; Chari, D.A.; Meek, R.; Rauch, S.D.; Remenschneider, A.K.; Quesnel, A.M.; de Venecia, R.; Lee, D.J.; Chien, W.; et al. Assessment of Sudden Sensorineural Hearing Loss After COVID-19 Vaccination. *JAMA Otolaryngol. Head Neck Surg.* **2022**, *148*, 307–315. [CrossRef] [PubMed]
26. Guo, W.; Deguise, J.; Tian, Y.; Huang, P.C.-E.; Goru, R.; Yang, Q.; Peng, S.; Zhang, L.; Zhao, L.; Xie, J.; et al. Profiling COVID-19 Vaccine Adverse Events by Statistical and Ontological Analysis of VAERS Case Reports. *Front. Pharmacol.* **2022**, *13*, 870599. [CrossRef] [PubMed]
27. Yanir, Y.; Doweck, I.; Shibli, R.; Najjar-Debbiny, R.; Saliba, W. Association Between the BNT162b2 Messenger RNA COVID-19 Vaccine and the Risk of Sudden Sensorineural Hearing Loss. *JAMA Otolaryngol. Head Neck Surg.* **2022**, *148*, 299–306. [CrossRef] [PubMed]
28. Nieminen, T.A.; Kivekäs, I.; Artama, M.; Nohynek, H.; Kujansivu, J.; Hovi, P. Sudden Hearing Loss Following Vaccination Against COVID-19. *JAMA Otolaryngol. Head Neck Surg.* **2023**, *149*, 133–140. [CrossRef] [PubMed]
29. Thai-Van, H.; Valnet-Rabier, M.B.; Anciaux, M.; Lambert, A.; Maurier, A.; Cottin, J.; Pietri, T.; Destère, A.; Damin-Pernik, M.; Perrouin, F.; et al. Safety Signal Generation for Sudden Sensorineural Hearing Loss Following Messenger RNA COVID-19 Vaccination: Postmarketing Surveillance Using the French Pharmacovigilance Spontaneous Reporting Database. *JMIR Public Health Surveill.* **2023**, *9*, e45263. [CrossRef]
30. Lin, R.J.; Krall, R.; Westerberg, B.D.; Chadha, N.K.; Chau, J.K. Systematic review and meta-analysis of the risk factors for sudden sensorineural hearing loss in adults. *Laryngoscope* **2012**, *122*, 624–635. [CrossRef]
31. Kahn, B.B.; Apostolidis, S.A.; Bhatt, V.; Greenplate, A.R.; Kallish, S.D.; LaCava, A.; Lucas, A.B.; Meyer, N.J.; Negoianu, D.; Ogdie, A.R.M.; et al. Multisystem Inflammation and Organ Dysfunction After BNT162b2 Messenger RNA Coronavirus Disease 2019 Vaccination. *Crit. Care Explor.* **2021**, *3*, e0578. [CrossRef]
32. Pisani, D.; Leopardi, G.; Viola, P.; Scarpa, A.; Ricciardiello, F.; Cerchiai, N.; Astorina, A.; Chiarella, G. Sudden sensorineural hearing loss after covid-19 vaccine; A possible adverse reaction? *Otolaryngol. Case Rep.* **2021**, *21*, 100384. [CrossRef]

vaccines

MDPI

Article

Evaluation of Antibody Response and Adverse Effects following Heterologous COVID-19 Vaccine Booster with mRNA Vaccine among Healthcare Workers in Indonesia

Gatot Soegiarto [1,2,*], Bagus Aulia Mahdi [3], Laksmi Wulandari [4], Karin Dhia Fahmita [3], Satrio Tri Hadmoko [3], Hendra Ikhwan Gautama [3], Dewi Prasetyaningtyas [3], Muhammad Edwin Prasetyo [3], Pujo Prawiro Negoro [3], Nur Arafah [3], Dewajani Purnomosari [5], Damayanti Tinduh [6], Dominicus Husada [7], Ari Baskoro [1], Deasy Fetarayani [1], Wita Kartika Nurani [3] and Delvac Oceandy [8,9,*]

1 Division of Allergy and Clinical Immunology, Department of Internal Medicine, Faculty of Medicine, Universitas Airlangga—Dr. Soetomo General Academic Hospital, Surabaya 60132, Indonesia; ari.baskoro@fk.unair.ac.id (A.B.); deasy-f@fk.unair.ac.id (D.F.)
2 Postgraduate School, Master Program on Immunology, Universitas Airlangga, Surabaya 60132, Indonesia
3 Department of Internal Medicine, Faculty of Medicine, Universitas Airlangga—Dr. Soetomo General Academic Hospital, Surabaya 60132, Indonesia; bagus.aulia.mahdi-2018@fk.unair.ac.id (B.A.M.); karin.dhia.fahmita-2017@fk.unair.ac.id (K.D.F.); satrio.tri.hadmoko-2017@fk.unair.ac.id (S.T.H.); hen.ikhwan.gautama-2017@fk.unair.ac.id (H.I.G.); dewi.prasetyaningtyas-2019@fk.unair.ac.id (D.P.); mohammad.edwin.prasetyo-2018@fk.unair.ac.id (M.E.P.); pujo.prawiro.negoro-2019@fk.unair.ac.id (P.P.N.); nur.arafah-2018@fk.unair.ac.id (N.A.); wita.kartika@fk.unair.ac.id (W.K.N.)
4 Department of Pulmonology and Respiratory Medicine, Faculty of Medicine, Universitas Airlangga—Dr. Soetomo General Academic Hospital, Surabaya 60132, Indonesia; laksmi.wulandari@fk.unair.ac.id
5 Department of Histology and Cell Biology, Faculty of Medicine, Public Health and Nursing, Universitas Gadjah Mada, Yogyakarta 55281, Indonesia; d.purnomosari@ugm.ac.id
6 Department of Physical Medicine and Rehabilitation, Faculty of Medicine, Universitas Airlangga—Dr. Soetomo General Academic Hospital, Surabaya 60132, Indonesia; damayanti.tinduh@fk.unair.ac.id
7 Department of Child Health, Faculty of Medicine, Universitas Airlangga—Dr. Soetomo General Academic Hospital, Surabaya 60132, Indonesia; dominicus.husada@fk.unair.ac.id
8 Division of Cardiovascular Sciences, Faculty of Biology Medicine and Health, Manchester Academic Health Science Centre, University of Manchester, Manchester M13 9PT, UK
9 Department of Biomedical Science, Faculty of Medicine, Universitas Airlangga, Surabaya 60132, Indonesia
* Correspondence: gatot_soegiarto@fk.unair.ac.id (G.S.); delvac.oceandy@manchester.ac.uk (D.O.)

Abstract: *Background*: The administration of the third (or booster) dose of COVID-19 vaccine is important in maintaining protection against SARS-CoV-2 infection or the severity of the disease. In Indonesia, health care workers (HCWs) are among the first to receive a booster dose of the COVID-19 vaccine. In this study, we evaluated the antibody response and adverse events following heterologous booster vaccine using mRNA-1273 among HCWs that were fully vaccinated with inactivated viral vaccine as the priming doses. *Methods*: 75 HCWs at Dr. Soetomo General Hospital in Surabaya, Indonesia, participated in this study. The level of antibody against the SARS-CoV-2 receptor binding domain was analyzed at 1, 3, and 5 months following the second priming dose and at 1, 3, and 5 months after the booster dose. *Results*: We found a significantly higher level of antibody response in subjects receiving a booster dose of the mRNA-1273 vaccine compared to those receiving an inactivated viral vaccine as a booster. Interestingly, participants with hypertension and a history of diabetes mellitus showed a lower antibody response following the booster dose. There was a higher frequency of adverse events following injection with the mRNA-1273 vaccine compared to the inactivated viral vaccine, although the overall adverse events were considered minor. *Conclusions*: A heterologous booster dose using mRNA vaccine resulted in a high antibody response; however, participants with hypertension and diabetes mellitus displayed a lower antibody response.

Keywords: COVID-19; vaccine; booster; comorbidity; antibody response; healthcare worker; adverse event; mRNA vaccine; inactivated viral vaccine

Citation: Soegiarto, G.; Mahdi, B.A.; Wulandari, L.; Fahmita, K.D.; Hadmoko, S.T.; Gautama, H.I.; Prasetyaningtyas, D.; Prasetyo, M.E.; Negoro, P.P.; Arafah, N.; et al. Evaluation of Antibody Response and Adverse Effects following Heterologous COVID-19 Vaccine Booster with mRNA Vaccine among Healthcare Workers in Indonesia. *Vaccines* **2023**, *11*, 1160. https://doi.org/10.3390/vaccines11071160

Academic Editor: Kay Choong See

Received: 18 April 2023
Revised: 19 June 2023
Accepted: 23 June 2023
Published: 26 June 2023

1. Introduction

The national vaccination program for Coronavirus Disease 2019 (COVID-19) in Indonesia started on 13 January 2021 with health care workers (HCWs) being one of the first groups of people receiving the vaccine. The success of the vaccination program among HCWs is crucial to protecting them from the risk of contracting COVID-19 from their work. Recent reports have shown a considerable high acceptance of the COVID-19 vaccine in people from low- and middle-income countries (LMIC) [1]. However, other studies have demonstrated a variability among HCWs in Asia and Africa in terms of willingness to take the COVID-19 vaccine [2–4]. Our own observation of HCWs in two major hospitals in East Java, Indonesia, indicated a high uptake of COVID-19 vaccination, i.e., more than 80% have received two doses of vaccine within the first 3 months after vaccine roll-out [5].

The inactivated viral vaccine was the main type of vaccine used by the Indonesian authorities at the beginning of vaccination. One of the inactivated SARS-CoV-2 vaccines, CoronaVac, which was used in the national vaccination program, showed a very good level of protection against severe COVID-19, hospitalization, and mortality [6,7]. However, recent published data have demonstrated a waning of antibody levels and protection following COVID-19 vaccination over time [8–11]. Observations on people who were vaccinated with the mRNA vaccine have suggested a decrease in serum antibody levels by 38% in each following month [9]. This might lead to an increase in the incidence of COVID-19 breakthrough infections, as reported elsewhere [11]. Despite a number of observations on mRNA vaccines, there were fewer reports regarding waning immunity and the reduction of antibody levels in people vaccinated with inactivated viral vaccines.

The importance of the COVID-19 vaccine booster is evident. It has been demonstrated that a third (booster) dose enhanced both humoral and cellular immunity regardless of the type of the priming doses and the type of the booster dose itself [12]. Moreover, the side effects of booster doses seem acceptable, as shown by the relatively non-severe adverse effects in people receiving booster vaccines, as reported in a previous publication [12].

We have evaluated the serum antibody levels against SARS-CoV-2 among HCWs in East Java, Indonesia, following vaccination using inactivated virus. We observed a significant increase in the serum antibody level [13]. However, we discovered that participants with hypertension displayed lower serum antibody levels compared to those with normal blood pressure [13].

In this present study we performed a follow up analysis of the same cohort of HCWs as described above, who received a booster dose of SARS-CoV-2 vaccine at 5 months following the priming dose. Most of the participants received the mRNA-1273 vaccine (Moderna) as a booster dose following recommendations from the Indonesian Health Authority, while a few of them opted to receive a booster dose using an inactivated viral vaccine (CoronaVac). Here we assessed the antibody response, the adverse effects of the vaccine, and the association of the antibody response with hypertension following the third (booster) dose of vaccines in our cohort of HCWs.

2. Materials and Methods

2.1. Study Participants

This is a follow-up observation on our previously reported study [13]. We recruited non-pregnant health care workers (HCWs) at Dr. Soetomo General Hospital in Surabaya, Indonesia. Apart from chronic underlying conditions such as diabetes, hypertension, and allergic diseases, the participants did not have any other diseases at the beginning of the study. All of the participants were tested for the presence of antibodies against SARS-CoV-2 before the start of the study. Participants with a detectable level of serum IgG against SARS-CoV-2 RBD before the first dose of vaccination and those who contracted COVID-19 (confirmed by PCR test) during the course of the study were excluded from the study since any infection with SARS-CoV-2 virus before or during the study might affect the serum antibody level against SARS-CoV-2 virus and hence confound the data. From 101 individuals who were originally involved in the study, 8 had positive serum IgG against SARS-CoV-2 before the priming dose of vaccination, 18 contracted SARS-CoV-2 infection during the course of the study, and 2 dropped out due to unwillingness to undergo follow-up examination. Thus, 75 participants fulfilled the criteria and were included in the analysis.

2.2. COVID-19 Vaccination

All participants received two doses of the inactivated SARS-CoV-2 vaccine (CoronaVac) as the priming doses. An analysis of serum samples was conducted at 1, 3, and 5 months after the second dose of vaccination. All participants were offered a booster dose with either the mRNA-1273 vaccine (Moderna) or the inactivated viral vaccine (CoronaVac) at 5–6 months following the second dose. Serum samples were again analyzed at 1, 3, and 5 months after the booster dose.

2.3. Serology Assay

We used a commercially available kit (Elecsys Anti-SARS-CoV-2 S, Roche Diagnostics, Mannheim, Germany) to examine the level of IgG against the SARS-CoV-2 receptor-binding domain (RBD) in the serum samples. We followed the protocol as recommended by the manufacturer.

2.4. Demographic and Adverse Events Data Collection

At the beginning of the study, we interviewed participants regarding demographic data and the presence or history of comorbidities (i.e., diabetes mellitus, cardiovascular disease, and allergic disease). Blood pressure was also measured at the beginning of the study and during follow-up visits. During the follow-up visits, participants were also asked about the presence of vaccine adverse reactions.

2.5. Statistical Analysis

The serum IgG level is presented as geometric mean titres and 95% confidence intervals (CI). To analyze the difference in IgG level between booster vaccination with mRNA vs. inactivated viral vaccine, we used a non-parametric multiple comparisons (Mann–Whitney U) test. The same test was also used to analyze the effects of comorbidities on the antibody response. A p value less than 0.05 was considered statistically significant. We used GraphPad Prism ver. 9 (GraphPad Software, LLC, Boston, MA, USA) to analyze the data. To control for the possible confounding effects of each comorbidity, a multivariate logistic regression analysis was performed with inclusion of histories of hypertension, diabetes mellitus, and cardiovascular diseases.

3. Results

3.1. Study Participants

This is a follow-up observation on our previously reported study [13]. However, in the present analysis, we only included HCWs who have never been infected with SARS-CoV-2 and have an undetectable level of serum IgG against the SARS-CoV-2 receptor binding domain (RBD) before the first dose of vaccination. Participants who contracted COVID-19 (confirmed by PCR test) during the course of the study and those who had baseline IgG levels against SARS-CoV-2 RBD were excluded from the study. A total of 75 HCWs who fulfilled these criteria were included in the analysis. The mean age of participants was 50.95 years old, and 60% of them were male. Hypertension was detected in 29.3% of participants. Some of the participants have a history of diabetes mellitus (21.3%), cardiovascular diseases (14.7%), and allergic diseases (42.7%) (Table 1).

Table 1. Characteristics of study participants.

	Participants Included in This Study (n = 75)
Sex	
Male	45 (60%)
Female	30 (40%)
Age at vaccination (y)	
Mean ± SD	50.95 ± 19.55
Median	57
Blood pressure	
Non-hypertension	53 (70.7%)
Hypertension (BP ≥ 140/90)	22 (29.3%)
History of diabetes mellitus	
No	59 (78.7%)
Yes	16 (21.3%)
History of cardiovascular diseases	
No	64 (85.3%)
Yes	11 (14.7%)
History of allergic diseases	
No	43 (57.3%)
Yes	32 (42.7%)

SD—standard deviation; BP—blood pressure.

3.2. Serum IgG Level against SARS-CoV-2 Receptor Binding Domain (RBD)

All of the participants received inactivated viral vaccine (CoronaVac) as the priming doses (first and second doses). Of the 75 HCWs in our cohort, 69 of them received the mRNA-1273 vaccine (Moderna) as a booster (third dose), whereas 6 subjects opted to have an inactivated viral vaccine (CoronaVac) for the booster dose. We analyzed serum IgG levels against the RBD domain at 1, 3, and 5 months following the 2nd dose of priming vaccine and at 1, 3, and 5 months after receiving the booster dose. As shown in Figure 1, we observed a marked increase in serum IgG levels following priming doses. Interestingly, the level of serum IgG did not significantly decrease up to 5 months after the priming doses. A marked increase in antibody levels was observed in all participants following the booster dose. The level of serum IgG against SARS-CoV-2 seemed higher in participants receiving the mRNA-1273 vaccine compared to those who had an inactivated viral vaccine as a booster (Figure 1). Although the difference reached statistical significance, it needs to be interpreted cautiously since the number of subjects receiving boosters with inactivated vaccines was very low.

3.3. Antibody Response in Participants with Comorbidities

Our previous observation has suggested that subjects with hypertension display a lower antibody response against the priming doses of inactivated viral vaccine [13]. To understand if hypertension affected the response to the booster dose using the mRNA-1273 vaccine, we compared serum IgG levels between participants with hypertension (BP ≥ 140/90) and those without hypertension. Interestingly, we found a consistent finding that participants with high BP showed a lower antibody response following booster doses using the mRNA-1273 vaccine (Figure 2A).

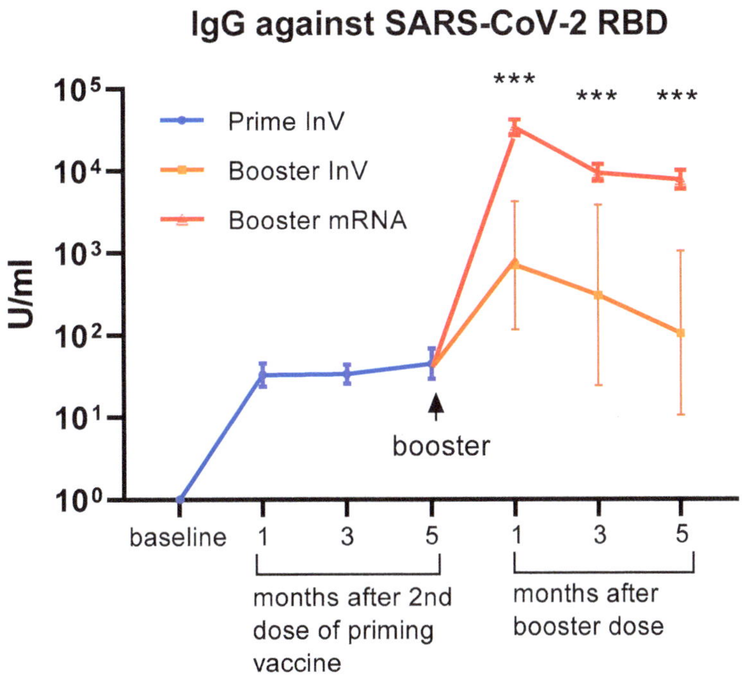

Figure 1. Serum antibody levels (IgG) against the SARS-CoV-2 receptor binding domain (RBD) at 1–5 months following priming (1st and 2nd doses) vaccination and at 1–5 months after the booster dose of COVID-19 vaccination. All of the participants (n = 75) received an inactivated viral vaccine for the priming dose. A total of 69 participants received the mRNA-1273 vaccine as a booster, whereas 6 participants received an inactivated viral vaccine as a booster. The data are presented as GMT and 95% CI. *** $p < 0.001$, multiple non-parametric test (Mann–Whitney U test), InV = inactivated viral vaccine.

The antibody response in participants with histories of diabetes mellitus (DM) and allergic diseases was also analyzed. Similar to participants with hypertension, we observed significantly lower antibody levels in subjects with a history of diabetes mellitus at 1–5 months following booster vaccination (Figure 2B). However, in contrast with the finding above, participants with allergic diseases displayed comparable levels of serum IgG following boosters compared to those without allergic diseases (Figure 2C).

To analyze the possible confounding effect between these three comorbidities, we conducted a multivariate linear regression analysis. Data presented in Tables 2–4 suggested that hypertension showed the strongest association with serum IgG level post-booster vaccination and remained significantly associated with antibody response, in particular at 1 month and 3 months after booster vaccination (Tables 2 and 3). History of diabetes mellitus showed a trend of significance at 1 month post-booster vaccination, whereas there was no significant association between history of allergic disease and serum IgG level. Overall, our finding showed that among co-morbidities, hypertension significantly influences the antibody response following a booster dose of the mRNA vaccine.

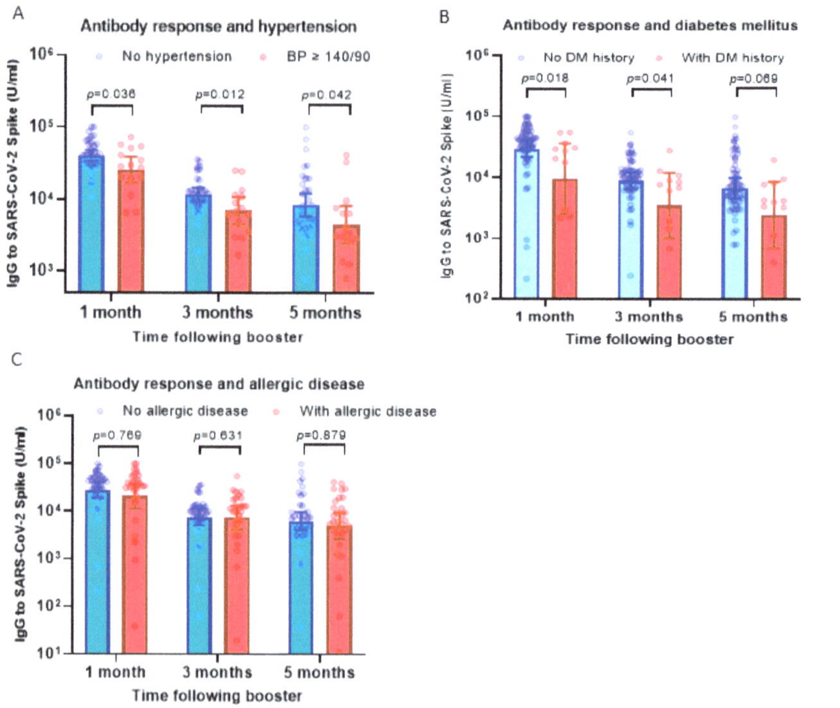

Figure 2. Antibody response following booster dose in participants with comorbidities. (**A**) Serum IgG levels against SARS-CoV-2 RBD in participants with hypertension (blood pressure/BP ≥ 140/90) compared to subjects with normal blood pressure. On average, the IgG levels of participants with hypertension were 30–43% lower compared to subjects without hypertension. (**B**) IgG levels in participants with a history of diabetes mellitus. On average, the IgG levels of participants with diabetes mellitus were 40–60% lower compared to subjects without hypertension. (**C**) IgG levels in participants with a history of allergic diseases. The serum IgG levels were comparable between subjects with and without allergic diseases. Multiple non-parametric tests (the Mann–Whitney U test) were used to compare the differences.

Table 2. Multivariate regression analysis of serum IgG level at 1 month post-booster dose.

Variable	Regression Coefficient	*p* Value
Hypertension	**−0.235**	**0.05**
History off Diabetes Mellitus	−0.232	0.07
History of allergic disease	0.006	0.96

Variable with significant association is written in bold.

Table 3. Multivariate regression analysis of serum IgG level at 3 months post-booster dose.

Variable	Regression Coefficient	*p* Value
Hypertension	**−0.246**	**0.05**
History off Diabetes Mellitus	−0.185	0.16
History of allergic disease	0.114	0.38

Variable with significant association is written in bold.

Table 4. Multivariate regression analysis of serum IgG level at 5 months post-booster dose.

Variable	Regression Coefficient	*p* Value
Hypertension	−0.168	0.18
History off Diabetes Mellitus	−0.139	0.28
History of allergic disease	−0.060	0.64

3.4. Adverse Reactions following Vaccination

The pattern of the adverse effects is depicted in Figure 3A,B. In total, there were 26 events of adverse reactions reported after first dose of priming vaccination, 18 reactions following the second dose of priming vaccine, and 48 reactions after the booster dose of vaccination. It is clear that booster injections using the mRNA-1273 vaccine triggered more frequent adverse effects than the priming doses using inactivated viral vaccine. The most frequent adverse effect following a booster dose was pain at the injection site, which was reported by 20 participants (25.4%), followed by fever (10 participants, 12.7%) and muscle pain (6 participants, 6.3%). The adverse effects of the first and second priming doses using inactivated viral vaccine were less frequent. For example, only 10–12% reported local pain following injection, whereas other side effects occurred in less than 5% of the participants. However, despite the higher frequency of adverse effects following booster vaccination, the majority of them were non-severe and transient. None of the participants required hospital treatment due to the side effects of the vaccine.

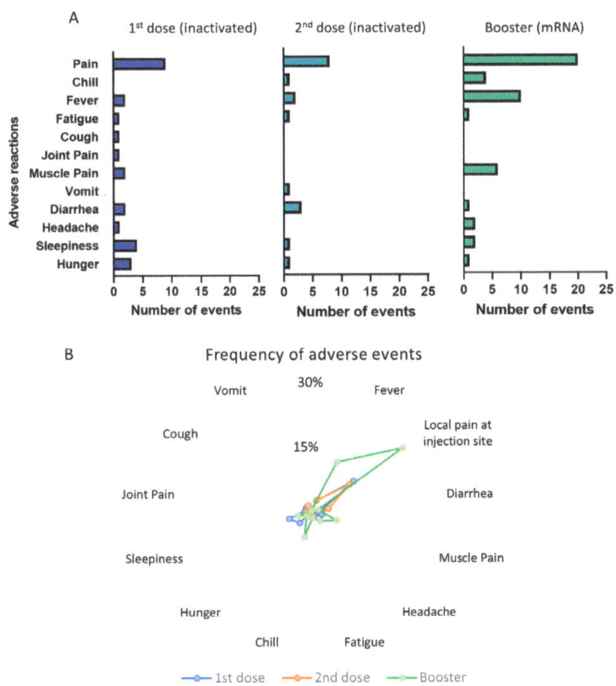

Figure 3. Adverse events following priming and booster doses of COVID-19 vaccination. (**A**) Number of adverse events following the 1st and 2nd doses of inactivated viral vaccine and booster doses of mRNA vaccine. (**B**) Radial graph showing the frequency of adverse events comparing the 1st, 2nd, and booster doses of vaccination.

4. Discussion

The main finding of this study is that a heterologous booster dose of the mRNA-1273 vaccine induces a strong antibody response in individuals who have been vaccinated with an inactivated viral vaccine. This data is in line with previous reports and adds to the growing body of evidence showing the effectiveness of heterologous booster vaccination in enhancing antibody levels against SARS-CoV-2 [12,14–16]. Another important finding of our study is that participants with hypertension and a history of diabetes mellitus (DM) exhibited a lower antibody response following the booster dose compared to those with normal blood pressure or without DM history.

The effectiveness of mRNA vaccines as booster doses in enhancing antibody titers is evident. When compared to other platforms of COVID-19 vaccine, for example, adenovirus, recombinant protein, and inactivated viral vaccine, the mRNA vaccine showed superiority in terms of the level of antibody response [12]. Based on a study by Zhang et al., it was stated that the long-term antibody levels following vaccination with inactivated viral vaccine at 11–12 months post-vaccination are very low. Consequently, people who were vaccinated with an inactivated viral vaccine might still be susceptible to SARS-CoV-2 infection, although they have a lower severity of COVID-19 infection than those who were not vaccinated [17]. Our present data is in line with previous reports indicating a strong antibody response to the mRNA vaccine as a booster, regardless of the type of the priming vaccine. Importantly, we also found a significantly enhanced antibody response in participants who received inactivated viral vaccine as a booster dose, although it was at a significantly lower level than the response against mRNA booster vaccine. However, it is important to note that the number of subjects in this category was very small.

One novel finding of this study is that the antibody response against the booster dose is significantly lower in participants with hypertension (BP $\geq$ 140/90) and those with a history of DM compared to subjects without these comorbidities. Previous observations have indicated a reduction in antibody response following the COVID-19 vaccine in subjects with hypertension [13,18–22] and diabetes mellitus [21–26]. These phenomena were reported on subjects who received inactivated viral vaccines [13,18], mRNA vaccines [19,20,22,24] and adenoviral vaccine [25] as the priming doses. Our present data indicate that (i) subjects with hypertension and diabetes mellitus also showed reduced antibody responses following the third or booster dose, and (ii) the reduction of the response occurred following heterologous booster vaccination with mRNA vaccine. This further supports the idea that hypertension and diabetes mellitus may play a very important role in determining antibody responses against vaccination. Indeed, the association between hypertension, diabetes mellitus, and dysregulation of immune response has been widely documented previously [27,28], and our study has added to the line of evidence in terms of vaccine response. Further studies to delineate the precise mechanism are needed in order to better understand this phenomenon and find strategies to minimize the detrimental effect of high blood pressure in reducing vaccine response.

It is important to note that the median age of the participants was relatively high (57 years). Interestingly, the proportion of participants with hypertension in our cohort was lower compared to the prevalence of hypertension in the Indonesian population within the age group of 55–64 years old [29]. However, the prevalence of diabetes in our cohort seemed comparable to the prevalence in the general population at a similar age (19.6%) [29].

The other important finding of this study was that the antibody response to the booster dose declined over time. Our analysis suggested that the antibody titer at 5 months after the booster dose declined by more than 70% compared to the antibody titer at 1 month post-booster. Evidence showing the waning immunity following priming doses of COVID vaccine is accumulating [8,10], and this has become one of the main reasons for the importance of having a booster dose. Our data provide new evidence that the serum antibody level is also declining following a booster dose. We do not have evidence whether this decline will result in a reduction in protection since we did not assess the incidence of breakthrough infections in our cohort. The status of the cellular immunity, which is associated with protection against disease severity, also needs to be evaluated to understand if the protection following a booster dose is also reduced over time. However, it is important to note that recent observations suggested that people with a lower antibody response were more prone to breakthrough infections against the new Omicron variants [30].

In addition to antibody response, another important aspect that needs to be assessed is the effectiveness of controlling the incidence and severity of COVID-19. In our previous observation, we analyzed the incidence and severity of COVID-19 among a cohort of HCWs who received vaccination with an inactivated viral vaccine [5]. We assessed COVID-19 incidence and severity by comparing the period before and after the start of the national vaccination program. We found a significant reduction in COVID-19 hospitalizations in the period after vaccination compared to before vaccination. However, we still observed a higher incidence of infection in the fully vaccinated cohort at the time when a new variant circulated in the population [5]. The incidence of breakthrough SARS-CoV-2 infection due to a new variant (in the case of our cohort, the Delta variant) indicated that an appropriate strategy for booster vaccination is necessary to control COVID-19 incidence and severity in the future. Equally important, the use of multivalent COVID-19 vaccines should always be considered given the occurrence of multiple SARS-CoV-2 variants in Indonesia, as indicated by data in the GISAID database [31].

The mRNA vaccine has been widely associated with more side effects compared to other types of vaccines [12,32]. Our data concur with previous studies, in which we found more frequent adverse effects, both systemic and local, following vaccination using mRNA vaccine compared to the effects after vaccination with inactivated viral vaccine. However, it is important to note that most of the adverse effects in our cohort were temporary, and none of them required hospitalization. This indicates that the booster vaccination using mRNA vaccine is generally safe for people who have received inactivated viral vaccine before.

Our present study has several limitations. First, we only examined humoral responses by assessing serum IgG levels. As mentioned previously, cellular immunity is important in protecting against disease severity. Further studies are needed to understand whether cellular immunity is enhanced following a heterologous primary vaccine-booster combination of inactivated viral vaccine with mRNA vaccine and whether cellular immunity is waning over time following booster. Second, we only have a small number of participants in this study. It is important to note that a number of study subjects were excluded due to infection with the SARS-CoV-2 virus during the study period. We recruited infection-naïve subjects, and so some of the participants who had detectable antibodies at the beginning of the study were excluded. Another limitation was the possibility of asymptomatic SARS-CoV-2 infection among the participants following the priming doses, which might affect the level of serum IgG antibody. We did not perform an analysis of anti-nucleocapsid antibodies, which could detect asymptomatic infection, due to limited resources. However, we closely monitored for the presence of COVID-19 symptoms among all the participants and followed this up with a PCR test during the study, so the possibility of asymptomatic SAS-CoV-2 infection was probably minimal. The detection of anti-nucleocapsid antibody may also be useful in comparing antibody responses against inactivated virus vs. mRNA vaccines, since inactivated viral vaccines may contain N-proteins in addition to S-proteins, whereas mRNA vaccines only contain S-antigens. Studies to focus on this aspect are needed in the future.

It is clear that HCWs need to be protected against COVID-19 since they may have higher exposure to the virus due to possible contacts with patients. In our recent observation, we found a higher incidence of COVID-19 in medical staff (i.e., physicians and nurses) compared to non-medical hospital workers, such as administrative staff, during the first year of the COVID-19 pandemic in 2020 [5]. This is in line with other reports that showed a higher risk of healthcare workers contracting the SARS-CoV-2 infection [33–36]. This underlines the importance of giving healthcare workers appropriate protection, including protective clothing and booster doses of vaccines.

In summary, our findings show that heterologous booster doses using mRNA vaccine in a cohort of HCWs who received priming vaccine with inactivated virus induce a higher antibody response than homologous booster vaccine with inactivated SARS-CoV2. However, the level of serum IgG waned over time and participants with hypertension displayed a lower antibody response. Our data adds to the growing body of evidence showing the importance of the COVID-19 booster vaccination to improve protection against the disease.

Author Contributions: Conceptualization, G.S. and D.O.; Data curation, D.F.; Formal analysis, G.S. and D.O.; Funding acquisition, L.W.; Investigation, B.A.M., K.D.F., S.T.H., H.I.G., D.P. (Dewi Prasetyaningtyas), M.E.P., P.P.N. and N.A.; Methodology, D.T.; Project administration, W.K.N.; Resources, A.B.; Software, D.H.; Supervision, D.T.; Validation, G.S., L.W., D.P. (Dewajani Purnomosari) and D.O.; Visualization, D.O.; Writing—original draft, G.S. and D.O.; Writing—review and editing, G.S., D.P. (Dewajani Purnomosari) and D.O. All authors have read and agreed to the published version of the manuscript.

Funding: This research was funded by Universitas Airlangga Mandate Research Grant, grant number 1043/UN3.15/PT/2021. The APC was funded by Universitas Airlangga.

Institutional Review Board Statement: This study has been reviewed and has obtained ethical approval from the Local Health Research Ethics Committee of Dr. Soetomo General Academic Hospital, Surabaya, Indonesia (No. 0145/KEPK/II/2021).

Informed Consent Statement: Informed consent has been obtained from all participants involved in the study.

Data Availability Statement: The datasets created and analyzed during this study are available from the corresponding authors upon reasonable request.

Acknowledgments: The authors would like to thank Anisa Octaviani and Astri Nur Amalia for administrative support.

Conflicts of Interest: The authors declare no conflict of interest.

References

1. Solis Arce, J.S.; Warren, S.S.; Meriggi, N.F.; Scacco, A.; McMurry, N.; Voors, M.; Syunyaev, G.; Malik, A.A.; Aboutajdine, S.; Adeojo, O.; et al. COVID-19 vaccine acceptance and hesitancy in low- and middle-income countries. *Nat. Med.* **2021**, *27*, 1385–1394. [CrossRef] [PubMed]
2. Alhassan, R.K.; Owusu-Agyei, S.; Ansah, E.K.; Gyapong, M. COVID-19 vaccine uptake among health care workers in Ghana: A case for targeted vaccine deployment campaigns in the global south. *Hum. Resour. Health* **2021**, *19*, 136. [CrossRef]
3. Elhadi, M.; Alsoufi, A.; Alhadi, A.; Hmeida, A.; Alshareea, E.; Dokali, M.; Abodabos, S.; Alsadiq, O.; Abdelkabir, M.; Ashini, A.; et al. Knowledge, attitude, and acceptance of healthcare workers and the public regarding the COVID-19 vaccine: A cross-sectional study. *BMC Public Health* **2021**, *21*, 955. [CrossRef]
4. Wong, E.L.; Qiu, H.; Chien, W.T.; Wong, J.C.; Chalise, H.N.; Hoang, H.T.; Nguyen, H.T.; Chan, P.K.; Wong, M.C.; Cheung, A.W.; et al. COVID-19 Vaccine Willingness and Related Factors among Health Care Workers in 3 Southeast Asian Jurisdictions. *JAMA Netw. Open* **2022**, *5*, e2228061. [CrossRef] [PubMed]
5. Soegiarto, G.; Purnomosari, D.; Wulandari, L.; Aulia Mahdi, B.; Dhia Fahmita, K.; Tri Hadmoko, S.; Ikhwan Gautama, H.; Edwin Prasetyo, M.; Prasetyaningtyas, D.; Prawiro Negoro, P.; et al. Incidence of SARS-CoV-2 infection in hospital workers before and after vaccination programme in East Java Indonesia—A retrospective cohort study. *Lancet Reg. Health Southeast Asia* **2023**, *10*, 100130. [CrossRef]
6. Jara, A.; Undurraga, E.A.; Gonzalez, C.; Paredes, F.; Fontecilla, T.; Jara, G.; Pizarro, A.; Acevedo, J.; Leo, K.; Leon, F.; et al. Effectiveness of an Inactivated SARS-CoV-2 Vaccine in Chile. *N. Engl. J. Med.* **2021**, *385*, 875–884. [CrossRef] [PubMed]
7. Tanriover, M.D.; Doganay, H.L.; Akova, M.; Guner, H.R.; Azap, A.; Akhan, S.; Kose, S.; Erdinc, F.S.; Akalin, E.H.; Tabak, O.F.; et al. Efficacy and safety of an inactivated whole-virion SARS-CoV-2 vaccine (CoronaVac): Interim results of a double-blind, randomised, placebo-controlled, phase 3 trial in Turkey. *Lancet* **2021**, *398*, 213–222. [CrossRef]
8. Chemaitelly, H.; Tang, P.; Hasan, M.R.; AlMukdad, S.; Yassine, H.M.; Benslimane, F.M.; Al Khatib, H.A.; Coyle, P.; Ayoub, H.H.; Al Kanaani, Z.; et al. Waning of BNT162b2 Vaccine Protection against SARS-CoV-2 Infection in Qatar. *N. Engl. J. Med.* **2021**, *385*, e83. [CrossRef] [PubMed]
9. Israel, A.; Shenhar, Y.; Green, I.; Merzon, E.; Golan-Cohen, A.; Schaffer, A.A.; Ruppin, E.; Vinker, S.; Magen, E. Large-Scale Study of Antibody Titer Decay following BNT162b2 mRNA Vaccine or SARS-CoV-2 Infection. *Vaccines* **2021**, *10*, 64. [CrossRef]
10. Levin, E.G.; Lustig, Y.; Cohen, C.; Fluss, R.; Indenbaum, V.; Amit, S.; Doolman, R.; Asraf, K.; Mendelson, E.; Ziv, A.; et al. Waning Immune Humoral Response to BNT162b2 COVID-19 Vaccine over 6 Months. *N. Engl. J. Med.* **2021**, *385*, e84. [CrossRef]

11. Lin, D.Y.; Gu, Y.; Wheeler, B.; Young, H.; Holloway, S.; Sunny, S.K.; Moore, Z.; Zeng, D. Effectiveness of COVID-19 Vaccines over a 9-Month Period in North Carolina. *N. Engl. J. Med.* **2022**, *386*, 933–941. [CrossRef] [PubMed]
12. Munro, A.P.S.; Janani, L.; Cornelius, V.; Aley, P.K.; Babbage, G.; Baxter, D.; Bula, M.; Cathie, K.; Chatterjee, K.; Dodd, K.; et al. Safety and immunogenicity of seven COVID-19 vaccines as a third dose (booster) following two doses of ChAdOx1 nCov-19 or BNT162b2 in the UK (COV-BOOST): A blinded, multicentre, randomised, controlled, phase 2 trial. *Lancet* **2021**, *398*, 2258–2276. [CrossRef]
13. Soegiarto, G.; Wulandari, L.; Purnomosari, D.; Dhia Fahmita, K.; Ikhwan Gautama, H.; Tri Hadmoko, S.; Edwin Prasetyo, M.; Aulia Mahdi, B.; Arafah, N.; Prasetyaningtyas, D.; et al. Hypertension is associated with antibody response and breakthrough infection in health care workers following vaccination with inactivated SARS-CoV-2. *Vaccine* **2022**, *40*, 4046–4056. [CrossRef]
14. Lu, B.N.; Zhu, K.L.; Cui, X.M.; Yao, L.; Wang, X.J.; Wang, G.L.; Duan, L.J.; Qian, A.; Ma, M.J. Antibody and T-cellular response to COVID-19 booster vaccine in SARS-CoV-1 survivors. *Clin. Immunol.* **2022**, *244*, 109103. [CrossRef]
15. Silva, M.F.S.; Pinto, A.; Oliveira, F.C.E.; Caetano, L.F.; Araujo, F.M.C.; Fonseca, M.H.G. Antibody response six months after the booster dose of Pfizer in previous recipients of CoronaVac. *J. Med. Virol.* **2022**, *95*, e28169. [CrossRef]
16. Yigit, M.; Ozkaya-Parlakay, A.; Cosgun, Y.; Ince, Y.E.; Kalayci, F.; Yilmaz, N.; Senel, E. Antibody response after a booster dose of BNT162B2mRNA and inactivated COVID-19 vaccine. *J. Clin. Virol. Plus* **2022**, *2*, 100094. [CrossRef]
17. Zhang, S.; Xu, K.; Li, C.; Zhou, L.; Kong, X.; Peng, J.; Zhu, F.; Bao, C.; Jin, H.; Gao, Q.; et al. Long-Term Kinetics of SARS-CoV-2 Antibodies and Impact of Inactivated Vaccine on SARS-CoV-2 Antibodies based on a COVID-19 Patients Cohort. *Front. Immunol.* **2022**, *13*, 829665. [CrossRef]
18. Dinc, H.O.; Saltoglu, N.; Can, G.; Balkan, I.I.; Budak, B.; Ozbey, D.; Caglar, B.; Karaali, R.; Mete, B.; Tuyji Tok, Y.; et al. Inactive SARS-CoV-2 vaccine generates high antibody responses in healthcare workers with and without prior infection. *Vaccine* **2022**, *40*, 52–58. [CrossRef] [PubMed]
19. Ebinger, J.E.; Joung, S.; Liu, Y.; Wu, M.; Weber, B.; Claggett, B.; Botting, P.G.; Sun, N.; Driver, M.; Kao, Y.H.; et al. Demographic and clinical characteristics associated with variations in antibody response to BNT162b2 COVID-19 vaccination among healthcare workers at an academic medical centre: A longitudinal cohort analysis. *BMJ Open* **2022**, *12*, e059994. [CrossRef]
20. Fujigaki, H.; Yamamoto, Y.; Koseki, T.; Banno, S.; Ando, T.; Ito, H.; Fujita, T.; Naruse, H.; Hata, T.; Moriyama, S.; et al. Antibody Responses to BNT162b2 Vaccination in Japan: Monitoring Vaccine Efficacy by Measuring IgG Antibodies against the Receptor-Binding Domain of SARS-CoV-2. *Microbiol. Spectr.* **2022**, *10*, e0118121. [CrossRef]
21. Kung, Y.A.; Huang, S.Y.; Huang, C.G.; Liu, K.T.; Huang, P.N.; Yu, K.Y.; Yang, S.L.; Chen, C.P.; Cheng, C.Y.; Lee, I.K.; et al. Factors influencing neutralizing antibody titers elicited by coronavirus disease 2019 vaccines. *Microbes Infect.* **2022**, *25*, 105044. [CrossRef] [PubMed]
22. Mitsunaga, T.; Ohtaki, Y.; Seki, Y.; Yoshioka, M.; Mori, H.; Suzuka, M.; Mashiko, S.; Takeda, S.; Mashiko, K. The evaluation of factors affecting antibody response after administration of the BNT162b2 vaccine: A prospective study in Japan. *PeerJ* **2021**, *9*, e12316. [CrossRef]
23. Boroumand, A.B.; Forouhi, M.; Karimi, F.; Moghadam, A.S.; Naeini, L.G.; Kokabian, P.; Naderi, D. Immunogenicity of COVID-19 vaccines in patients with diabetes mellitus: A systematic review. *Front. Immunol.* **2022**, *13*, 940357. [CrossRef] [PubMed]
24. Priddy, F.H.; Williams, M.; Carson, S.; Lavender, B.; Mathieson, J.; Frampton, C.; Moreland, N.J.; McGregor, R.; Williams, G.; Brewerton, M.; et al. Immunogenicity of BNT162b2 COVID-19 vaccine in New Zealand adults. *Vaccine* **2022**, *40*, 5050–5059. [CrossRef]
25. Rangsrisaeneepitak, V.; Porntharukchareon, T.; Dechates, B.; Sirisreetreerux, S.; Tawinprai, K. Antibody levels in people with diabetes after one dose of the ChAdOx1 nCoV-19 (AZD1222) vaccine. *Diabetol. Int.* **2022**, *13*, 637–643. [CrossRef]
26. Virgilio, E.; Trevisan, C.; Abbatecola, A.; Malara, A.; Palmieri, A.; Fedele, G.; Stefanelli, P.; Leone, P.; Schiavoni, I.; Maggi, S.; et al. Diabetes Affects Antibody Response to SARS-CoV-2 Vaccination in Older Residents of Long-Term Care Facilities: Data from the GeroCovid Vax Study. *Diabetes Care* **2022**, *45*, 2935–2942. [CrossRef] [PubMed]
27. Daryabor, G.; Atashzar, M.R.; Kabelitz, D.; Meri, S.; Kalantar, K. The Effects of Type 2 Diabetes Mellitus on Organ Metabolism and the Immune System. *Front. Immunol.* **2020**, *11*, 1582. [CrossRef]
28. Singh, M.V.; Chapleau, M.W.; Harwani, S.C.; Abboud, F.M. The immune system and hypertension. *Immunol. Res.* **2014**, *59*, 243–253. [CrossRef]
29. Badan Penelitian dan Pengembangan Kesehatan, Kementrian Kesehatan RI—Ministry of Health the Republic of Indonesia. *Laporan Nasional RISKESDAS—National Report on Basic Health Survey*; Kementrian Kesehatan RI—Ministry of Health the Republic of Indonesia: Jakarta, Indonesia, 2018.
30. Mohlendick, B.; Ciuciulkaite, I.; Elsner, C.; Anastasiou, O.E.; Trilling, M.; Wagner, B.; Zwanziger, D.; Jockel, K.H.; Dittmer, U.; Siffert, W. Individuals with Weaker Antibody Responses after Booster Immunization Are Prone to Omicron Breakthrough Infections. *Front. Immunol.* **2022**, *13*, 907343. [CrossRef]
31. Khare, S.; Gurry, C.; Freitas, L.; Schultz, M.B.; Bach, G.; Diallo, A.; Akite, N.; Ho, J.; Lee, R.T.; Yeo, W.; et al. GISAID's Role in Pandemic Response. *China CDC Wkly.* **2021**, *3*, 1049–1051. [CrossRef] [PubMed]
32. Romero-Ibarguengoitia, M.E.; Gonzalez-Cantu, A.; Pozzi, C.; Levi, R.; Mollura, M.; Sarti, R.; Sanz-Sanchez, M.A.; Rivera-Salinas, D.; Hernandez-Ruiz, Y.G.; Armendariz-Vazquez, A.G.; et al. Analysis of immunization time, amplitude, and adverse events of seven different vaccines against SARS-CoV-2 across four different countries. *Front. Immunol.* **2022**, *13*, 894277. [CrossRef] [PubMed]

33. Doernberg, S.B.; Holubar, M.; Jain, V.; Weng, Y.; Lu, D.; Bollyky, J.B.; Sample, H.; Huang, B.; Craik, C.S.; Desai, M.; et al. Incidence and prevalence of COVID-19 within a healthcare worker cohort during the first year of the SARS-CoV-2 pandemic. *Clin. Infect. Dis.* **2022**, *75*, 1573–1584. [CrossRef] [PubMed]
34. Inghels, M.; Kane, R.; Lall, P.; Nelson, D.; Nanyonjo, A.; Asghar, Z.; Ward, D.; McCranor, T.; Kavanagh, T.; Hogue, T.; et al. Ethnicity and risk for SARS-CoV-2 infection among the healthcare workforce: Results of a retrospective cohort study in rural United Kingdom. *Int. J. Infect. Dis.* **2022**, *122*, 115–122. [CrossRef] [PubMed]
35. Platten, M.; Nienhaus, A.; Peters, C.; Cranen, R.; Wisplinghoff, H.; Kersten, J.F.; Bach, A.D.; Michels, G. Cumulative Incidence of SARS-CoV-2 in Healthcare Workers at a General Hospital in Germany during the Pandemic-A Longitudinal Analysis. *Int. J. Environ. Res. Public Health* **2022**, *19*, 2429. [CrossRef] [PubMed]
36. Sabetian, G.; Moghadami, M.; Hashemizadeh Fard Haghighi, L.; Shahriarirad, R.; Fallahi, M.J.; Asmarian, N.; Moeini, Y.S. COVID-19 infection among healthcare workers: A cross-sectional study in southwest Iran. *Virol. J.* **2021**, *18*, 58. [CrossRef]

vaccines

MDPI

Article

Evaluating the Adverse Events Associated with Three Doses of the COVID-19 Vaccination in Adults in the Western Region of Saudi Arabia: A Cross-Sectional Study

Hamzah J. Aldali [1,*], Jehad A. Aldali [2], Badi A. Alotaibi [3,4], Glowi A. Alasiri [5], Aroob M. Alromih [6], Emadeldin M. Elsokkary [7], Ali Z. Aldali [8] and Abdullah Almeziny [6]

1 Cellular and Molecular Medicine, College of Biomedical Science, University of Bristol, Bristol BS8 1TD, UK
2 0921 Department of Pathology, Faculty of Medicine, Imam Mohammad Ibn Saud Islamic University, Riyadh 13317, Saudi Arabia
3 Department of Clinical Laboratory Sciences, Collage of Applied Medical Sciences, King Saud Bin Abdulaziz University for Health Sciences, Riyadh 11481, Saudi Arabia
4 King Abdullah International Medical Research Center, Riyadh P.O. Box 3660, Saudi Arabia
5 Department of Biochemistry, College of Medicine Organization Imam Mohammad Ibn Saud Islamic University, Riyadh 13317, Saudi Arabia
6 Medical School, Imam Mohammad Ibn Saud Islamic University, Riyadh 13317, Saudi Arabia
7 Psychology, Organisation Imam Mohammad Ibn Saud Islamic University, Riyadh 13317, Saudi Arabia
8 Department of Rehabilitation Health Science—Physical Therapy, College of Applied Medical Sciences, King Saud University, Riyadh 11445, Saudi Arabia
* Correspondence: ng21108@bristol.ac.uk

Abstract: The Kingdom of Saudi Arabia was one of the countries earliest affected by the coronavirus 2019 (COVID-19) pandemic and had taken precautions including compulsory COVID-19 vaccination. Both the ChAdOx1 nCoV-19 vaccine (Oxford AstraZeneca) and the BNT162b2 vaccine (Pfizer) were approved by the Saudi Ministry of Health, followed by mRNA-1273 (Moderna), all of which were used for population-wide vaccination. This study aimed to assess the short-term side effects following the COVID-19 vaccinations among participants who had received all three doses in the western region of Saudi Arabia. An online survey was distributed to the participants who received either BNT162b2, ChAdOx1 nCoV-19, or mRNA-1273 vaccines, and the type of side effects and their severity were evaluated. Fatigue and headache, pain at the site of the injection and muscle pain were the most common side effects in all three doses. However, the severity depending on the type of vaccination was significant only for the first and second dose, but not the third dose. In contrast, there was a higher percentage of participants who encountered severe side effects from the third dose compared to the first and second. Nevertheless, the majority of participants described all three doses' side effects to be moderately severe. A future evaluation could be made to access the individual types of vaccination and compare between the side effects of the BNT162b2, ChAdOx1 nCoV-19, and mRNA-1273 vaccines specifically for the booster dose.

Keywords: vaccines; COVID-19; coronavirus; infectious diseases; adverse events

Citation: Aldali, H.J.; Aldali, J.A.; Alotaibi, B.A.; Alasiri, G.A.; Alromih, A.M.; Elsokkary, E.M.; Aldali, A.Z.; Almeziny, A. Evaluating the Adverse Events Associated with Three Doses of the COVID-19 Vaccination in Adults in the Western Region of Saudi Arabia: A Cross-Sectional Study. *Vaccines* **2023**, *11*, 266. https://doi.org/10.3390/vaccines11020266

Academic Editor: Kay Choong See

Received: 24 December 2022
Revised: 23 January 2023
Accepted: 24 January 2023
Published: 26 January 2023

1. Introduction

Despite many lockdowns and long-term infection control measures implemented in most countries, the 2019 coronavirus (COVID-19) pandemic, which started in China in December 2019, is still spreading [1–3]. Since the beginning of the COVID-19 pandemic, both hospital and community infection control efforts have been employed to minimize the risk of infection spread and some safe vaccines have been generated [4]. Several vaccination options have been accessible for use and found to be safe and effective. By the end of 2020, many countries, including Gulf countries, employed vaccination campaigns due to the promising effectiveness of the COVID-19 vaccines. However, all of the vaccines proposed by the World Health Organization do not provide complete immunization against the disease [4,5]. Therefore, administering more than one dose was proposed in many countries [6]. Comparing COVID-19 infections with seasonal influenza, it was found that the influenza vaccine contributed to lower rates of hospitalization and mortality, while the COVID-19 vaccination significantly reduced death and hospitalization rates in the elderly group only [7]. COVID-19 vaccination campaigns have been introduced to reach the public and private sectors in many countries. Furthermore, some countries made it mandatory to be vaccinated in order to enter the country [8]. The Kingdom of Saudi Arabia was one of the first countries to take precautionary actions including compulsory COVID-19 vaccination for the first, followed by the second and third dosage of vaccination [9,10]. More importantly, to lessen the impact of the disease, Saudi health officials implemented early, unprecedented preventive measures and precautionary strategies, including the cancellation of social events, prohibiting international flights, closing schools and universities, work-from-home mandates, curfews, and placing the entire nation under complete lockdown [5,9]. These approaches were incredibly effective. These initiatives did not include reliable test and contact tracing protocols. As a result, a second wave of the virus hit numerous Arab Gulf nations, prompting additional lockdowns [5]. However, in the absence of an approved antiviral treatment for COVID-19, several vaccine development studies were quickly launched in hopes of bringing the pandemic under control [6]. As a result, immunization programs started in developed countries and then expanded to other parts of the world, but the administrative problems of manufacturing, transporting, and managing billions of doses on a global scale presented unprecedented challenges, and these challenges had to be addressed while assessing the short-term and long-term effects of the vaccines [5]. The adoption of vaccines has been demonstrated to be challenging, and vaccinations alone have been shown to be insufficient to shift the pandemic from its acute to its chronic phase. Continuing unprecedented preventive measures and precautionary strategies should be included alongside vaccination campaigns, especially to avoid the long-term spread of the virus and the emergence of new variants [5]. In early 2021, different international health authorities declared various vaccines to have emergency use authorization [9,11,12]. The first vaccines that were introduced and approved in the Kingdom of Saudi Arabia were the ChAdOx1 nCoV-19 (also known as Oxford AstraZeneca) vaccine and the BNT162b2 (also known as Pfizer) vaccine [6]. The Food and Drug Authority approved the two vaccines. BNT162b2 is a nucleic acid vaccine based on a modified mRNA molecule that encodes for the spike protein of SARS-CoV-2. ChAdOx1 nCoV-19 is a modified viral-vector based vaccine derived from the chimpanzee adenovirus, ChAdOx1, and encodes the SARS-CoV-2 spike protein, BNT162b2ChAdOx1 nCoV-19 [13].

On 16 July 2021, the mRNA-1273 (also known as Moderna) vaccine was added to the list of approved vaccines in the Kingdom of Saudi Arabia [14]. Several mild to moderate side effects, such as headaches, pain, swelling, and redness at the injection site, as well as muscle and joint aches, were associated with the side effects of the COVID-19 vaccinations [9,15,16]. In the eastern region of Saudi Arabia, similar common side effects during the first and second dose were reported such as fatigue, headache, and fever. Unusual side effects that have been previously reported include palpitations and irregular menstruation [14,17]. Further side effects such as pain at the injection site, feeling tired, and headaches were reported in Riyadh (the capital city of Saudi Arabia) [18].

There is limited literature regarding the third dose vaccination's side effects in the general population [19], and in the population of Saudi Arabia [20], in particular, the western region of Saudi Arabia. Describing the consequences following the third vaccination dose, also known as the booster vaccination, compared with the first and second vaccination doses will aid in enhancing the understanding of the safety profile of the third dose compared to the first and second dose of the COVID-19 vaccinations and improve the vaccination process against COVID-19 [21,22]. Thus, this study aims to evaluate the short-term adverse effects of the third COVID-19 vaccination dose compared with the first and second vaccination doses in the Kingdom of Saudi Arabia among participants in the western region.

2. Materials and Methods

An online survey was distributed using a Google form with dual language (Arabic and English) for a cross-sectional study and was distributed to participants who were vaccinated with the BNT162b2, ChAdOx1 nCoV-19 and mRNA-1273 vaccines. The side effects were reported following the participants' vaccinations. The survey was distributed in the western region of the Kingdom of Saudi Arabia (mainly in Makkah, Taif, and Jeddah) between the period of 26 January and 8 March 2022. The survey was revised by all authors to provide feedback on the survey sections and recommend any edits if needed.

The online questionnaire was designed and distributed on social media platforms including WhatsApp and Twitter; emails were also circulated to public health and university staff. Following entry to the online questionnaire, participants were asked to carefully read the comprehensive explanation of the purpose of the study prior to giving consent on a compulsory electronic consent form, comprising the data for voluntary participation and anonymity. An e-mail address was generated to facilitate communication between the participants and study researchers. Upon completion of the online survey, the data was anonymously collected and stored in a safe file. The structured online questionnaire contained two sections: The first section aimed to explore the participants' demographical information (gender, age, nationality, and education) alongside their SARS-CoV-2 infection status (chronic conditions, previous infection with COVID-19, consumed medications and antibiotics). The second section aimed to investigate the type of vaccine received, the side effects post vaccination (first, second, or third dose), the duration of the encountered side effects, and any analgesics consumed to reduce the severity of the side effects. In addition, on days 1, 2, 3, and 4 following immunizations, the participants were requested to illustrate the intensity degree of each symptom, ranging from mild to severe. Additionally, the participants were asked about the average timing of the onset of side effects. All the participants who declined to take part, or who were not vaccinated with three doses, and participants who received vaccines other than BNT162b2, -ChAdOx1 nCoV-19, or mRNA-1273 were excluded. The sample size was calculated through raosoft.com, which indicated 385 participants as a sufficient sample size to achieve a 5.35% margin of error and 95% confidence.

2.1. Statistical Analysis

For the data, the statistical analysis was performed using Statistics Package Social Science (SPSS) Version 25 (IBM Armonk, New York, NY, USA). Descriptive statistics were used as qualitative data with a p-value ≤ 0.05 was considered significant. The descriptive statistics were expressed as qualitative data to compare the symptoms between the types of vaccines and their side effect. Using the Mann–Whitney U test, the researchers conducted dimensional comparisons between the groups described herein.

2.2. Ethical Approval

The committee of the Institutional Review Board (IRB) at Imam Mohammad ibn Saud Islamic University has reviewed and approved this research with project number 167-2021, dated 20 December 2021. The IRB-approved study was titled, 'Evaluate the Side Effect Associated with Three Dosage of the Covid-19 Vaccine on Adults in western Region, Saudi Arabia: A Cross-Sectional Study'.

3. Results

3.1. Demographic Characteristics of the Participants and Medical History

In this study, 574 participants were involved, but 161 were eliminated, because they did not meet the inclusion criteria. Therefore, the final sample size was reduced to 413 participants. The majority of participants were between 18 and 25 years of age (n = 138,

33.4%), followed by 41 and 60 years (n = 119, 28.8%), then 31 and 40 years (n = 78, 18.8%), then 26 and 30 years (n = 70, 16.9%), and only eight (1.9%) participants were >60 years of age. In terms of nationalities, 344 (83.2%) were Saudi Arabian, while the remaining 69 (16.7%) were non-Saudi Arabian participants (Table 1). The applicants' general history was summarized in Table 1, with 84.3% of participants declaring no clinical history, while 44 (10.7%) reported a chronic disease and 21 (5.1%) reported previous health problems.

Table 1. Overview of participant demographics and medical history.

Variables	Patient Number	Percent (%)
Gender		
Male	262	63.4
Female	151	36.6
Age groups		
18–25	138	33.4
26–30	70	9.4
31–40	78	34.4
41–60	119	29.1
>60	8	3.5
Nationality		
Saudi	344	83.3
Non-Saudi	69	16.7
Education		
Teacher	136	33.4
Student	111	26.8
Non-student/Unemployed	90	21.7
Other employment	76	18.4
Health status		
Good health	348	84.3
Health problem	21	5.1
Chronic disease	44	10.7
Diagnosed with coronavirus infection before you received the first dose of the coronavirus vaccine	41	17.3
Not diagnosed with coronavirus infection before you received the first dose of the coronavirus vaccine	196	82.7
Receiving antimicrobial agents	6	1.2
Not receiving antimicrobial agents	407	98.5
Taking any medications to treat any disease	96	23.2
Not taking any medications to treat any disease	317	76.8

3.2. Participants' Side Effects per Dose

In Table 2, the side effects were reported for the first, second, and third doses of the COVID-19 vaccinations (ChAdOx1 nCoV-19, mRNA-1273, and BNT162b2).

First doses: Most participants received BNT162b2 BioNTech (71.4%), followed by Oxford-ChAdOx1 nCoV-19 (27.3%), and mRNA-1273 (0.7%) (Table 2). Nearly 38% of the participants experienced side effects the day after vaccination. In comparison, 43.6 % of participants showed adverse effects on the second and third days after vaccination. The most reported adverse effects among the trial participants were injection site pain (54%), followed by muscle and/or joint pain (36.3%), then fatigue and headache (35.1%). However, menstrual disorder, dizziness, vomiting, breathing congestion, chest pain, hair loss, and skin itching or rash were less commonly reported by the study participants (Tables 2 and 3). A total of 74% of participants indicated that the severity of the side effects was mild or moderately severe; leaving only 11.9% who suffered from severe adverse side effects. Most participants consumed pain relief medication to reduce the side effects' severity (64.9%). In contrast, only 16.9% of participants indicated that there were no adverse effects following the first vaccination.

Table 2. COVID-19 vaccination and the side effects encountered with BNT16b2, ChAdOx1 nCoV-19 and mRNA-1273.

	First Dose Number (%)	Second Dose Number (%)	Third Dose Number (%)
Type of vaccine			
BNT162b2	295 (71.4)	322 (78)	353 (85.5)
ChAdOx1 nCoV-19			
mRNA-1273	113 (27.3)	75 (18.2)	18 (4.4)
Diagnosed with coronavirus infection before	3 (0.7)	15 (3.5)	42 (10.1)
receiving the COVID-19 vaccination	41 (17.3)	0 (0)	46 (11.1)
Common side effects after vaccination			
Fatigue and/or headache	145 (35.1)	174 (42.1)	187 (45.3)
Pain at the site of injection	223 (54)	238 (57.6)	229 (55.4)
Muscle and/or joint pain	150 (36.3)	136 (32.9)	134 (32.4)
High temperature and shivering	56 (13.6)	174 (42.1)	118 (28.6)
Dizziness	223 (54)	238 (57.6)	60 (14.5)
Menstrual disorder	150 (36.3)	136 (32.9)	33 (8)
No side effects	138 (33.4)	48 (29.5)	139 (33.7)
Severity of side effects			
Mild	56 (13.6)	48 (11.6)	116 (28.1)
Moderate	51 (12.3)	44 (10.7)	157 (38)
Severe	69 (16.9)	70 (16.94)	89 (21.5)
When did you feel the side effects after the first dose of the coronavirus vaccine?			
First day	157 (38)	169 (40.9)	185 (44.8)
Second day	180 (43.6)	167 (40.4)	157 (38)
Third day	15 (3.6)	13 (3.1)	16 (3.9)
How long did the side effect last after vaccination?			
One—two days	226 (54.7)	243 (58.8)	224 (54.2)
Three days	97 (23.5)	77 (18.6)	77 (18.6)
Four days or more	36 (8.7)	34 (8.2)	56 (13.6)
Any medication taken?			
Medication taken to reduce the severity of the side effects.	268 (64.9)	257 (62.2)	258 (62.5)
No medication taken to reduce the severity of the side effects.	145 (35.1)	156 (37.8)	155 (37.5)

Table 3. Less common side effects after vaccine.

Less Common Side Effect	First Dose Number (%)	Second Dose Number (%)	Third Dose Number (%)
Vomiting	8 (1.9)	15 (3.6)	22 (5.3)
Breathing congestion	20 (4.8)	18 (4.4)	25 (6.1)
Skin itching or rash	20 (4.8)	21 (5.1)	25 (6.1)
Drop in sugar level	1 (0.2)	0 (0)	1 (0.2)
Chest pain	2 (0.5)	2 (0.5)	2 (0.5)

Second dose: Most participants received BNT162b2 (78%), followed by ChAdOx1 nCoV-19 (18.2%), and mRNA-1273 (3.6%) (Table 2). Nearly 40.9% of the participants experienced side effects on the next vaccination day. A total of 40.4% of participants showed adverse effects on the second and third days after vaccination. In particular, the most reported adverse effects among the trial participants were pain at the injection site (57.6%), fatigue and headache (42.1%), and muscle and/or joint pain (32.9%). However, dizziness, menstrual disorder, vomiting, breathing problems, hair loss, chest pain, and skin rashes and itching were reported less frequently by study participants (Tables 2 and 3). Only 13.8% of participants encountered severe side effects. However, 62% of the participants consumed some medication to reduce the severity of the side effects. Only 16.9% of participants indicated that there were no adverse effects following the second vaccination.

Third dose: Most participants were vaccinated with BNT162b2 (85.5%), followed by mRNA-1273 (10.1%), and ChAdOx1 nCoV-19 (4.4%) (Table 2). Almost 44.8% of the participants experienced side effects on the day after vaccination. In comparison, 38% of participants showed adverse effects on the second and third days after vaccination. In particular, the most reported adverse effects among the trial participants were injection site pain (55.4%), fatigue and headache (45.3%), and muscle and/or joint pain (32.4%). Less commonly and rarely reported side effects were menstrual disorder, vomiting, breathing congestion, hair loss, chest pain, and skin itching or rash. A total of 62.5% of participants received some medication to reduce the severity of the side effects.

Table 4 shows the differences in the severe symptoms depending on the type of COVID-19 vaccination. In fact, it was found that there were high significant differences (p value < 0.01) between those who received the BNT162b2 vaccine and those who took the ChAdOx1 nCoV-19 vaccine and mRNA-1273 vaccine in terms of the severity of symptoms after the first and second doses (p value of 0.0001 and 0.006, respectively). While there were no significant differences in the severity of symptoms after the third dose (p value of 0.867) (Table 4).

Table 4. The significance of symptoms' severity depending on the type of vaccination for the three doses.

	Type of COVID-19 Vaccine	N	Mean Rank	Kruskal–Wallis H	p Value
Side effects' severity following the first dose	BNT162b2	295	192.57	17.746	0.0001
	ChAdOx1 nCoV-19	113	244.66		
	mRNA-1273	5	207.50		
	Total	413			
Side effects' severity following the second dose	BNT162b2	322	205.65	10.115	0.006
	ChAdOx1 nCoV-19	75	194.64		
	mRNA-1273	16	292.06		
	Total	413			
Side effects' severity following the third dose	BNT162b2	353	205.93	0.286	0.867
	ChAdOx1 nCoV-19	18	219.31		
	mRNA-1273	42	210.71		
	Total	413			

4. Discussion

Most countries have taken precautions to limit the spread of SARS-CoV-2 since the COVID-19 pandemic started in December 2019 [9,10]. Saudi Arabia was one of the earliest countries to initiate early immunization efforts after COVID-19 vaccination approval by the Saudi Ministry of Health and the World Health Organization [9]. At the start of 2021, a number of vaccine candidates were authorized for emergency use by several international health organizations [9,11,12,23]. Initially, Saudi Arabia authorized the use of BNT162b2, followed by the vaccine developed by Oxford AstraZeneca, ChAdOx1 nCoV-19 [9], then the mRNA-1273 vaccine [14]. Consequently, in this study, participants who received three doses of the COVID-19 vaccine in the western Saudi Arabian region were evaluated for any short-term adverse events of the third COVID-19 vaccination dose compared to the first and second vaccination doses. In this study, most participants received BNT162b2, followed by ChAdOx1 nCoV-19 vaccines. Nevertheless, it was found that more participants encountered severe side effects with ChAdOx1 nCoV-19 for the first and second doses compare to BNT162b2-BioNTech and mRNA-1273. Alghamdi et al. also demonstrated that the severity of the side effects from ChAdOx1 nCoV-19 was higher compared to BNT162b2 in both the first and second doses. The incidence of mild adverse events was 30.1%, and 29.7% for severe side effects following the administration of the ChAdOx1 vaccination [14]. In comparison, the majority of participants who received BNT162b2 vaccinations only experienced mild side effects (63.92%); only 7.68% experienced severe side effects. However, our findings suggest that the severity of ChAdOx1 side effects were not significantly different compared with those for BNT162b2.

Patients who received the third COVID-19 vaccination dose encountered side effects such as pain at the site of injection, fatigue and headache, which were closely similar in percentage to the side effects following the second dose. Indeed, the most common systemic and local side effects of the COVID-19 vaccination were fatigue and tenderness [24]. It was found that the most common three side effects in all three dosages were fatigue and headache, pain at the injection site, and muscle and joint pain. These side effects were mostly common despite the difference in vaccination type. In contrast, the rare side effects reported included vomiting, breathing congestion, a skin rash, chest pain, and a drop in sugar level, showed very similar percentages between patients for all three doses. Supporting our findings, it was previously reported that the most common side effects included discomfort at the injection site, fever, headache, fatigue, and flu-like symptoms, while sleepiness, difficulty breathing, and body aches were less prevalent [25].

Our results showed that the most reported level of severity of the side effects was moderate. Unlike previous findings, a mild level of severity of symptoms was outstanding [14,21]. The majority of the findings concluded that the severity of the COVID-19 vaccinations ranged between moderate and mild, which supports the safety profile for the COVID-19 vaccines approved by the Saudi Arabian Ministry of Health [22]. Most of the participants reported that each dose's adverse effects lasted one—two days and took medication to decrease the severity of symptoms [25]. Similarly, the duration of symptoms after the first and second doses lasted for one—two days, which was highly reported previously [9,14,19].

The results demonstrated that there was a difference between genders in terms of the severity of symptoms for the first, second, and third doses, and the averages showed that the severity of symptoms was higher in females. This data supports previous research and indicates that females were considerably more likely than men to experience side effects following vaccination [14,15,20,26]. This might indicate that the Covid-19 vaccines function by stimulating the immune system, which can impact females more due to gender-based differences in the immune response, as observed in vaccines for diseases such as measles, mumps, and many others [24]. Furthermore, this may be attributable to hormonal variations. Since ACE2 expression is coded by the X chromosome, it is possible that males and females experience distinct patterns of its regulation [27]. In the past, it was revealed that women have lower levels of ACE2 expression in the lung, which led researchers to hypothesize that estrogen might decrease ACE2 expression [27]. Androgen also upregulates the mucosa-specific serine protease TMPRSS2, which helps viruses enter human host cells [28]. One of the enzymes that aids virus entry is TMPRSS2 [29]. This adds to the evidence that implies sex hormones may make it easier for males to become infected with COVID-19. Furthermore, several studies have shown that women develop stronger antibodies in response to infection and vaccination than men. For example, estrogens were shown to upregulate antibody development in mice, whereas testosterone suppressed antibody production. Furthermore, men with higher testosterone levels had a lower immune response to influenza vaccination than men with normal testosterone levels and women [30]. Several clinical trials have explored the application of hormone therapy to treat COVID-19 [31]. Because men are more likely to develop severe illness and die from the virus, researchers wondered if treating acute COVID-19 infection with female sex hormones could improve disease outcomes [6].

Furthermore, we found that there were statistically significant differences ($p < 0.001$) in the severity of symptoms between those who has taken analgesics to reduce the severity of their symptoms and those who did not after the three doses and found that the severity of the symptoms was greater in those who took analgesics.

Our study is the first Saudi Arabian western region-based investigation of COVID-19 vaccination adverse effects; however, this study had a few limitations. These include variations in patient interpretation and tolerance levels, and the fact that the outcomes of the questionnaire, which was circulated in the western region of Saudi Arabia, were self-reported by vaccinated participants and have not been clinically validated by expert clinicians. Heterogeneity in participant responses may have been caused by using a subjective scale rather than an objective standard to classify the severity of symptoms, such as mild, moderate, or severe. In addition, when completing the survey more than five days after vaccination, the participants were susceptible to recall bias, which hindered the accuracy of their memories. In addition, the survey did not question the incidence of immediate allergic reaction after vaccination for three doses. In order to comprehend the relationship between risk factors and developing side effects, larger participant studies need to be performed to widely evaluate the side-effects severity difference of the third dose compared to the first and second doses, depending on the type of vaccine administered. Moreover, many confounding variables may influence the interpretation of the data.

5. Conclusions

In conclusion, Saudi Arabia was one of the first countries to begin the administration of booster vaccinations for the whole Saudi Arabian population. Our cross-sectional study was concentrated in the western region of Saudi Arabia to evaluate the severity of side-effects for the third dose in comparison with the first and second vaccination doses. We found that the third dose had higher severe frequency of side effects as described by patients, which was not dependent on the type of immunization vaccine. In addition, there were no significant differences between the type of side effects reported for the three doses, including the rare side effects encountered in our study. These results indicate that the incidence of severe side effects following the third dose vaccination is frequent, while there are no differences between side effects for the first, second, and third doses after vaccination. Follow-up studies on larger populations are needed to assess vaccine efficacy in controlling and preventing COVID-19 infections, as well as the long- and short-term side effects.

Author Contributions: Conceptualization, J.A.A., H.J.A. and B.A.A.; data collection and analysis, E.M.E., A.Z.A., A.M.A. and A.A.; Review and editing, G.A.A.; Supervision, J.A.A. and H.J.A. All authors have read and agreed to the published version of the manuscript.

Funding: This research received no external funding.

Institutional Review Board Statement: The committee of the Institutional Review Board (IRB) at Imam Mohammad ibn Saud Islamic University has reviewed and approved this research with project number 167-2021, dated 20 December 2021.

Informed Consent Statement: Informed consent was obtained from all subjects involved in the study.

Data Availability Statement: The data used for the findings of the current study are available upon request from the corresponding author.

Conflicts of Interest: The authors declare no conflict of interest.

References

1. Jabal, K.A.; Ben-Amram, H.; Beiruti, K.; Batheesh, Y.; Sussan, C.; Zarka, S.; Edelstein, M. Impact of age, ethnicity, sex and prior infection status on immunogenicity following a single dose of the BNT162b2 mRNA COVID-19 vaccine: Real-world evidence from healthcare workers, Israel, December 2020 to January 2021. *Eurosurveillance* **2021**, *26*, 2100096. [CrossRef] [PubMed]
2. Alshammari, T.M.; Altebainawi, A.F.; Alenzi, K.A. Importance of early precautionary actions in avoiding the spread of COVID-19: Saudi Arabia as an Example. *Saudi Pharm. J.* **2020**, *28*, 898–902. [CrossRef] [PubMed]
3. Salam, A.A.; Al-Khraif, R.M.; Elsegaey, I. COVID-19 in Saudi Arabia: An overview. *Front. Public Health* **2022**, *9*, 2258. [CrossRef]
4. Shrotri, M.; Swinnen, T.; Kampmann, B.; Parker, E.P. An interactive website tracking COVID-19 vaccine development. *Lancet Glob. Health* **2021**, *9*, e590–e592. [CrossRef]

5. Althobaity, Y.; Wu, J.; Tildesley, M.J. Non-pharmaceutical interventions and their relevance in the COVID-19 vaccine rollout in Saudi Arabia and Arab Gulf countries. *Infect. Dis. Model.* **2022**, *7*, 545–560. [CrossRef]
6. Wiersinga, W.J.; Rhodes, A.; Cheng, A.C.; Peacock, S.J.; Prescott, H.C. Pathophysiology, transmission, diagnosis, and treatment of coronavirus disease 2019 (COVID-19): A review. *JAMA* **2020**, *324*, 782–793. [CrossRef]
7. Piroth, L.; Cottenet, J.; Mariet, A.-S.; Bonniaud, P.; Blot, M.; Tubert-Bitter, P.; Quantin, C. Comparison of the characteristics, morbidity, and mortality of COVID-19 and seasonal influenza: A nationwide, population-based retrospective cohort study. *Lancet Respir. Med.* **2021**, *9*, 251–259. [CrossRef] [PubMed]
8. Salvagno, G.L.; Henry, B.M.; Di Piazza, G.; Pighi, L.; De Nitto, S.; Bragantini, D.; Gianfilippi, G.L.; Lippi, G.; Giuseppe, S.k. Anti-SARS-CoV-2 receptor-binding domain total antibodies response in seropositive and seronegative healthcare workers undergoing COVID-19 mRNA BNT162b2 vaccination. *Diagnostics* **2021**, *11*, 832. [CrossRef]
9. Alhazmi, A.; Alamer, E.; Daws, D.; Hakami, M.; Darraj, M.; Abdelwahab, S.; Maghfuri, A.; Algaissi, A. Evaluation of side effects associated with COVID-19 vaccines in Saudi Arabia. *Vaccines* **2021**, *9*, 674. [CrossRef]
10. Algaissi, A.A.; Alharbi, N.K.; Hassanain, M.; Hashem, A.M. Preparedness and response to COVID-19 in Saudi Arabia: Building on MERS experience. *J. Infect. Public Health* **2020**, *13*, 834–838. [CrossRef]
11. Dal-Ré, R.; Caplan, A.L.; Gluud, C.; Porcher, R. Ethical and scientific considerations regarding the early approval and deployment of a COVID-19 vaccine. *Am. Coll. Physicians* **2021**, *174*, 258–260. [CrossRef]
12. Tumban, E. Lead SARS-CoV-2 candidate vaccines: Expectations from phase III trials and recommendations post-vaccine approval. *Viruses* **2020**, *13*, 54. [CrossRef] [PubMed]
13. Pedrazini, M.C.; da Silva, M.H. Pityriasis rosea-like cutaneous eruption as a possible dermatological manifestation after Oxford-AstraZeneca vaccine: Case report and brief literature review. *Dermatol. Ther.* **2021**, *34*, e15129. [CrossRef]
14. Alghamdi, A.N.; Alotaibi, M.I.; Alqahtani, A.S.; Al Aboud, D.; Abdel-Moneim, A.S. BNT162b2 and ChAdOx1 SARS-CoV-2 post-vaccination side-effects among saudi vaccinees. *Front. Med.* **2021**, *8*, 1796. [CrossRef]
15. Polack, F.P.; Thomas, S.J.; Kitchin, N.; Absalon, J.; Gurtman, A.; Lockhart, S.; Gruber, W.C. Safety and efficacy of the BNT162b2 mRNA COVID-19 vaccine. *N. Engl. J. Med.* **2020**, *383*, 2603–2615. [CrossRef] [PubMed]
16. Voysey, M.; Clemens, S.A.C.; Madhi, S.A.; Weckx, L.Y.; Folegatti, P.M.; Aley, P.K.; Angus, B.; Baillie, V.L.; Barnabas, S.L.; Bhorat, Q.E.; et al. Safety and efficacy of the ChAdOx1 nCoV-19 vaccine (AZD1222) against SARS-CoV-2: An interim analysis of four randomised controlled trials in Brazil, South Africa, and the UK. *Lancet* **2021**, *397*, 99–111. [CrossRef] [PubMed]
17. Farhat, M.; Al-Ibrahim, R.; Almohammedali, A.; Aljishi, R.; Alalwan, B. Study of the side effects of Pfizer and Oxford COVID-19 vaccines in the eastern province of Saudi Arabia. *Int. J. Gen. Med.* **2022**, *15*, 7547–7558. [CrossRef]
18. Aldali, J.A.; Alotaibi, F.T.; Alasiri, G.A.; Almesned, R.A.; Alromih, A.M.; Almohandes, A.M.; Alsenidi, S.F. Evaluate the side effect associated with COVID-19 vaccine on adolescents in Riyadh, Saudi Arabia. *Saudi Med. J.* **2022**, *43*, 1248–1253.
19. Patalon, T.; Saciuk, Y.; Peretz, A.; Perez, G.; Lurie, Y.; Maor, Y.; Gazit, S. Waning effectiveness of the third dose of the BNT162b2 mRNA COVID-19 vaccine. *Nat. Commun.* **2022**, *13*, 3203. [CrossRef]
20. Almughais, E.S.; Alharbi, A.H.; Aldarwish, H.A.; Alshammari, A.F.; Alsuhaymi, R.S.; Almuaili, J.A.; Alanizy, A.M. Side-effects of COVID-19 vaccines among the Saudi population. *Saudi Med. J.* **2022**, *43*, 386–393. [CrossRef]
21. Hatmal, M.m.M.; Al-Hatamleh, M.A.; Olaimat, A.N.; Hatmal, M.; Alhaj-Qasem, D.M.; Olaimat, T.M.; Mohamud, R. Side effects and perceptions following COVID-19 vaccination in Jordan: A randomized, cross-sectional study implementing machine learning for predicting severity of side effects. *Vaccines* **2021**, *9*, 556. [CrossRef]
22. Jayadevan, R.; Shenoy, R.S.; Anithadevi, T. Survey of symptoms following COVID-19 vaccination in India. *medRxiv* **2021**.
23. Ledford, H.; Cyranoski, D.; Van Noorden, R. The UK has approved a COVID vaccine—Here's what scientists now want to know. *Nature* **2020**, *588*, 205–206. [CrossRef]
24. Menni, C.; May, A.; Polidori, L.; Louca, P.; Wolf, J.; Capdevila, J.; Hu, C.; Ourselin, S.; Steves, C.J.; Valdes, A.M.; et al. COVID-19 vaccine waning and effectiveness and side-effects of boosters: A prospective community study from the ZOE COVID Study. *Lancet Infect. Dis.* **2022**, *22*, 1002–1010. [CrossRef]
25. El-Shitany, N.A.; Harakeh, S.; Badr-Eldin, S.M.; Bagher, A.M.; Eid, B.; Almukadi, H.; Alghamdi, B.; Alahmadi, A.; Hassan, N.A.; Sindi, N.; et al. Minor to moderate side effects of Pfizer-BioNTech COVID-19 vaccine among Saudi residents: A retrospective cross-sectional study. *Int. J. Gen. Med.* **2021**, *14*, 1389. [CrossRef] [PubMed]
26. Izumo, T.; Kuse, N.; Awano, N.; Tone, M.; Sakamoto, K.; Takada, K.; Muto, Y.; Fujimoto, K.; Saiki, A.; Itoet, Y.; et al. Side effects and antibody titer transition of the BNT162b2 messenger ribonucleic acid coronavirus disease 2019 vaccine in Japan. *Respir. Investig.* **2021**, *59*, 635–642. [CrossRef] [PubMed]
27. Gemmati, D.; Bramanti, B.; Serino, M.L.; Secchiero, P.; Zauli, G.; Tisato, V. COVID-19 and individual genetic susceptibility/receptivity: Role of ACE1/ACE2 genes, immunity, inflammation and coagulation. Might the double X-chromosome in females be protective against SARS-CoV-2 compared to the single X-chromosome in males? *Int. J. Mol. Sci.* **2020**, *21*, 3474. [CrossRef] [PubMed]
28. Lucas, J.M.; Heinlein, C.; Kim, T.; Hernandez, S.A.; Malik, M.S.; True, L.D.; Morrissey, C.; Corey, E.; Montgomery, B.; Mostaghel, E.; et al. The Androgen-Regulated Protease TMPRSS2 Activates a Proteolytic Cascade Involving Components of the Tumor Microenvironment and Promotes Prostate Cancer MetastasisTMPRSS2 Influences Prostate Cancer Metastasis. *Cancer Discov.* **2014**, *4*, 1310–1325. [CrossRef] [PubMed]

29. Hoffmann, M.; Kleine-Weber, H.; Schroeder, S.; Krüger, N.; Herrler, T.; Erichsen, S.; Schiergens, T.; Herrler, G.; Wu, N.-H.; Nitsche, A.; et al. SARS-CoV-2 cell entry depends on ACE2 and TMPRSS2 and is blocked by a clinically proven protease inhibitor. *Cell* **2020**, *181*, 271–280.e8. [CrossRef]
30. Tramontana, F.; Battisti, S.; Napoli, N.; Strollo, R. Immuno-endocrinology of COVID-19: The key role of sex hormones. *Front. Endocrinol.* **2021**, *12*, 726696. [CrossRef]
31. Choi, S.W.; Kim, J.; Lee, J.H.; Kim, S.K.; Lee, S.R.; Kim, S.H.; Chae, H.D. Hormone Therapy in the Era of the COVID-19 Pandemic: A Review. *J. Menopausal Med.* **2022**, *28*, 1–8. [CrossRef] [PubMed]

vaccines

MDPI

Article

Pandemic Fatigue and Vaccine Hesitancy among People Who Have Recovered from COVID-19 Infection in the Post-Pandemic Era: Cross-Sectional Study in China

Chenyuan Qin [1], Jie Deng [1], Min Du [1], Qiao Liu [1], Yaping Wang [1], Wenxin Yan [1], Min Liu [1] and Jue Liu [1,2,3,*]

1 School of Public Health, Peking University, Beijing 100191, China; qincy@bjmu.edu.cn (C.Q.); 1810306145@pku.edu.cn (J.D.); dumin@bjmu.edu.cn (M.D.); yaping77@bjmu.edu.cn (Y.W.); yanwx@bjmu.edu.cn (W.Y.); liumin@bjmu.edu.cn (M.L.)

2 Institute for Global Health and Development, Peking University, Beijing 100871, China

3 National Health Commission Key Laboratory of Reproductive Health, Peking University, Beijing 100191, China

* Correspondence: jueliu@bjmu.edu.cn; Tel.:+86-10-8-2801528-316; Fax:+86-10-8-2805146

Abstract: At present, the COVID-19 pandemic is still ongoing globally and the virus is constantly mutating. The herd immunity barrier established by past infections or vaccinations is gradually weakening and reinfections are occurring. To evaluate the pandemic fatigue and vaccine hesitancy among people who have recovered from COVID-19 in the post-pandemic era, we conducted an anonymous cross-sectional survey study in China from 4 July to 11 August 2023, nearly 6 months after the last large-scale nationwide infection. Basic sociodemographic characteristics, health-related factors (smoking, drinking, and chronic disease history), COVID-19 vaccination history, and self-reported long COVID were obtained as potential covariates. A series of logistic regression models were performed to examine the association between pandemic fatigue and vaccine hesitancy toward the next dose of COVID-19 vaccines via crude relative risks (cORs) and adjusted relative risks (aORs) with 95% CIs. According to our results, of the 2942 participants, 1242 (42.2%) were hesitant (unwilling or not sure) to receive the next dose of COVID-19 vaccines. The average score on the Pandemic Fatigue Scale was 21.67 ± 8.86, in which the scores of all items in the vaccine-hesitant group were significantly higher than those in the vaccine-accepting group. Additionally, the higher the pandemic fatigue level among people who have recovered from COVID-19, the more likely they were to be hesitant to receive the next dose of the COVID-19 vaccines (moderate: aOR = 2.94, 95% CI: 2.46–3.53; high: aOR = 6.88, 95% CI: 5.49–8.64). Overall, more than 40% of the recovered participants were unwilling or uncertain about the next vaccine dose, with varying degrees of pandemic fatigue. Pandemic fatigue is a potentially relevant factor for vaccine hesitancy and may hinder the translation of vaccination intention into behavior. Considering the ongoing reinfection situation, implementing a health education plan to reduce pandemic fatigue and prioritizing vaccination issues for people who have recovered from COVID-19 may be key to promoting the reduction of the COVID-19 disease burden and ensuring the health and well-being of the population.

Keywords: COVID-19; vaccination; pandemic fatigue; vaccine hesitancy

Citation: Qin, C.; Deng, J.; Du, M.; Liu, Q.; Wang, Y.; Yan, W.; Liu, M.; Liu, J. Pandemic Fatigue and Vaccine Hesitancy among People Who Have Recovered from COVID-19 Infection in the Post-Pandemic Era: Cross-Sectional Study in China. *Vaccines* **2023**, *11*, 1570. https://doi.org/10.3390/vaccines11101570

Academic Editor: Kay Choong See

Received: 23 August 2023
Revised: 29 September 2023
Accepted: 2 October 2023
Published: 5 October 2023

1. Introduction

The coronavirus disease 2019 (COVID-19) has been raging continuously for three years, indicating a great threat to human health [1]. As of 21 August 2023, there have been over 769.8 million confirmed cases and more than 6.9 million people have died worldwide [2]. Based on the comprehensive evaluation of facts including virus variation, epidemic situation, and the basis for previous prevention and control work, China's government successively introduced new optimization measures against COVID-19 in November

2022, trying to gradually bring people's lives back to normal [3,4]. From late 2022 to early 2023, China experienced a nationwide Omicron variant infection, and over 80% of the whole population was estimated to be infected [3,4]. Due to high asymptomatic infection rates and limited testing capabilities, the actual number of global infections may be much higher than the reported data [2]. On 5th May 2023, the Director-General of the World Health Organization (WHO) announced that COVID-19 no longer constitutes a public health emergency of international concern (PHEIC), lifting the highest-level alert issued on 30 January 2020 in Geneva [5]. However, the COVID-19 pandemic is not yet over and has become an ongoing global health problem [5]. The evolution of SARS-CoV-2 remains uncertain [5].

Vaccination, a vital weapon to actively prevent infectious diseases, is a milestone development in the history of human civilization [6,7]. Building a herd immunity barrier and minimizing infections, severe cases, and deaths through vaccination is an important guarantee for ensuring the normal operation of society under the current situation [6–8]. Vaccine hesitancy is defined as refusing or delaying vaccination when available [9]. As one of the top ten health threats announced by the WHO in 2019, vaccine hesitancy still cannot be ignored in the case of large-scale infections worldwide in 2023, especially since this acquired immunity is not permanent [9–11]. Since previous infections do not necessarily protect individuals from reinfection, it is essential for people who have recovered from COVID-19 to prevent reinfection by receiving the next dose of COVID-19 vaccines in time. However, we seem to have overlooked the vaccine hesitancy among people with a history of COVID-19 [11]. Early studies have shown that people who have recovered from COVID-19 were more likely to be unwilling or uncertain to receive the next dose of COVID-19 vaccines than those who have not been infected, but there is a lack of recent Chinese-specific research evidence [8,12,13]. Therefore, vaccine hesitancy among this group of people in China is an important issue that deserves early research.

As the 4-year mark is approaching since the commencement of the COVID-19 pandemic (November 2019), during which a "waxing and waning" battle trajectory has been observed almost globally and affected most aspects of human life, the concept of "pandemic fatigue" has become widely recognized in both academic and popular discourse [14]. Pandemic fatigue, defined by the WHO as "distress which can result in demotivation to follow recommended protective behaviors, emerging gradually over time and affected by several emotions, experiences, and perceptions", has been determined as an additional hurdle and risk factor to adherence to health-protective behavior by the public [15,16]. Originally developed by Lilleholt et al., the Pandemic Fatigue Scale (PFS) is a subjective questionnaire that has been used to measure subjects' fatigue from the COVID-19 pandemic as well as from the beginning of future pandemics [17]. Many factors affect vaccine hesitancy, such as basic demographic characteristics, trust in the government, risk perception of the pandemic, knowledge of COVID-19 and vaccines, and trust in the safety and efficacy of vaccines [11,18–21]. However, few studies are focusing on the attitudes towards the pandemic after a national epidemic and its impact on the vaccination intention among recovered people in China.

At present, the COVID-19 pandemic is still ongoing globally and the virus is constantly mutating [1]. Herd immunity established by infections or vaccinations is gradually weakening, and reinfections are occurring [9–11]. Therefore, it is essential to understand the status and correlates of pandemic fatigue and vaccine hesitancy toward the next dose among people who have been infected before. These findings will help government authorities and relevant parties to promote future vaccination policies and strategies in an orderly and precise manner, avoid complacency, and ensure the health and well-being of the population.

2. Methods

2.1. Study Design and Participants

To evaluate pandemic fatigue and vaccine hesitancy among people who have recovered from COVID-19 in the post-pandemic era, we conducted an anonymous cross-sectional survey in China from 4 July to 11 August 2023, nearly 6 months after the last large-scale nationwide SARS-CoV-2 infection [3,4]. This online survey was performed by a professional scientific data platform (Changsha Ranxing Information Technology Co., Ltd., Changsha, China) with nearly 300 million users every month [22]. It can accurately send the electronic questionnaire to our expected representative respondents based on the clear personal information (such as age, gender, and residence) of registered members [22]. Recruitment criteria were as follows: (1) agree to fill in the questionnaire carefully; (2) $\geq$18 years old; and (3) complete the survey for more than 300 s. Informed consent was embedded, and all respondents provided consent for anonymized data use for academic purposes.

Based on previous survey experience [19,20,23], to obtain sufficient participants with a history of COVID-19, we used a quota sampling method based on the population proportion of provinces reported in the *Seventh National Census* to allocate a total sample size of 3000 people. The predetermined sample size was obtained using a random sampling method in each province. After excluding unqualified replies and verifying the sufficient power of the test under α as 0.05 and the confidence interval width as 0.1 p (p = hesitancy rate in this study), 2942 participants with COVID-19 infection history were ultimately included in our analysis (Table S1).

2.2. Pandemic Fatigue

The Pandemic Fatigue Scale (PFS) is a subjective questionnaire that has been used as a tool to measure subjects' fatigue from the COVID-19 pandemic, which has been used cross-culturally with good validity and internal consistency [16,24]. After conducting translation into Chinese and back translation into English, we used an adapted PFS in Chinese in our questionnaire to ensure "linguistic and conceptual equivalence" [25]. The PFS comprises six items, grouped into two distinct yet highly correlated factors—behavioral and information fatigue—which both add to people's overall experience of pandemic fatigue [16]. Participants rank their agreement with each item on a 7-point Likert scale ranging from 1 (strongly disagree) to 7 (strongly agree) [16]. All items were combined with equal weighting, and the total score ranged from 6 to 42 points. Higher scores indicate more pronounced pandemic fatigue. For ease of statistical analysis, we divided pandemic fatigue into "Low (6–18 points)", "Moderate (19–30 points)", and "High (31–42 points)" levels based on the total scores. In the present study, the PFS exhibited acceptable internal consistency (α coefficient = 0.791).

2.3. Attitude toward the Next Dose of COVID-19 Vaccines

To determine the attitude of participants with a history of COVID-19 infection towards receiving the next dose of COVID-19 vaccines, we set a question as "Are you willing to receive the next dose of COVID-19 vaccine if available?". The answer was progressively set on a 5-point Likert scale (1 = very willing, 2 = willing, 3 = not sure, 4 = reluctant, 5 = very reluctant). We defined vaccine hesitancy as reluctance or uncertainty about receiving the next dose of COVID-19 vaccines, and then further asked for specific reasons for hesitancy.

2.4. Covariates

Sociodemographic characteristics (sex, age, location, education, relationship status), health-related factors (smoking, drinking, chronic disease history), and COVID-19 vaccination history were obtained as potential covariates. Risk perception items of COVID-19 reinfection were adapted from the Health Belief Model and involved two parts: perceived susceptibility and perceived severity [26]. Participants were asked "How likely do you think you are to be reinfected with COVID-19?" and "If you are reinfected with COVID-19, how severe do you think it will be?". The 5-point scale answers (1 = very low, 2 = low, 3 = moderate, 4 = high, 5 = very high) were divided into "Low", "Moderate", and "High" levels. In addition, we also included self-reported long COVID symptoms and set it as a binary variable (yes or no).

2.5. Statistics Analysis

Statistical descriptions of all quantitative variables were reported as frequencies, percentages, means (M), and standard deviations (SD). We used the Chi-square test and *t*-test to compare the group differences among basic characteristics of vaccine hesitancy and pandemic fatigue. A series of logistic regression models were performed to examine the association between pandemic fatigue and vaccine hesitancy via crude relative risks (cORs) and adjusted relative risks (aORs) with 95% CIs. Model A was unadjusted. We controlled sociodemographic factors (gender, age, location, and education) in model B, and then plus health factors (drinking and chronic disease history) in model C. Risk perception factors (perceived susceptibility and perceived severity) were additionally added to model D. Model E adjusted for all covariates that were significantly unequally distributed across three pandemic fatigue levels. On the basis of Model E, subgroup analyses were performed among participants with different characteristics. All statistical analyses were conducted by Software SPSS 26.0 (IBM SPSS Inc., New York, NY, USA), and we set the significance level at a two-sided p value of <0.05.

3. Results

3.1. Characteristics of Participants

Of the 2942 participants who have recovered from COVID-19 (Table 1), 1185 (40.3%) were male, 2180 (74.1%) were ≤34 years old, 2662 (90.5%) lived in urban areas, and 2178 (74%) had at least a bachelor's degree. Among them, a vast majority of participants were current non-smokers and only 33.6% struggled with chronic diseases. It is worth noting that less than 10% of respondents received a COVID-19 vaccination in the past six months, and 2150 (73.1%) of participants self-reported long COVID symptoms. In addition, 46.2% of all respondents believed that they had a low risk of reinfection, and more than 90% of them believed that the consequences of reinfection would not be severe.

Table 1. Characteristics and COVID-19 vaccine hesitancy among 2942 participants recovered from COVID-19 in China.

Characteristics †	Number (%)	Hesitancy Toward the Next Dose of COVID-19 Vaccine		
		n (%)	95% CI	*p* Value
Total	2942 (100)	1242 (42.2)	40.4–44.0	
Sex				0.06
Male	1185 (40.3)	525 (44.3)	41.5–47.1	
Female	1757 (59.7)	717 (40.8)	38.5–43.1	
Age (years)				0.502
<30	1390 (47.2)	575 (41.4)	38.8–44.0	
30–34	790 (26.9)	337 (42.7)	39.2–46.1	
35–39	421 (14.3)	174 (41.3)	36.7–46.1	
≥40	341 (11.6)	156 (45.7)	40.5–51.1	
Location				0.629
Urban	2662 (90.5)	1120 (42.1)	40.2–44.0	
Rural	280 (9.5)	122 (43.6)	37.9–49.4	
Education				0.960
High school and below	765 (26.0)	324 (42.4)	38.9–45.9	
Bachelor's degree	1928 (65.5)	811 (42.1)	39.9–44.3	
Master's degree	249 (8.5)	107 (43.0)	36.9–49.2	
Relationship status				0.001 *
Without partner	627 (21.3)	301 (48.0)	44.1–51.9	
With partner	2315 (78.7)	941 (40.6)	38.7–42.7	
Smoking				0.019 *
No	2493 (84.7)	1075 (43.1)	41.2–45.1	
Yes	449 (15.3)	167 (37.2)	32.8–41.7	

Table 1. *Cont.*

Characteristics †	Number (%)	Hesitancy Toward the Next Dose of COVID-19 Vaccine		
		n (%)	95% CI	*p* Value
Drinking				0.488
No	1660 (56.4)	710 (42.8)	40.4–45.2	
Yes	1282 (43.6)	532 (41.5)	38.8–44.2	
Chronic disease				0.880
No	1954 (66.4)	823 (42.1)	39.9–44.3	
Yes	988 (33.6)	419 (42.4)	39.4–45.5	
Perceived susceptibility				0.056
Low	1358 (46.2)	540 (39.8)	37.2–42.4	
Moderate	1204 (40.9)	531 (44.1)	41.3–46.9	
High	380 (12.9)	171 (45.0)	40.1–50.0	
Perceived severity				0.038 *
Low	1657 (56.3)	706 (42.6)	40.2–45.0	
Moderate	1000 (34.0)	439 (43.9)	40.8–47.0	
High	285 (9.7)	97 (34.0)	28.7–39.7	
Time of the most recent vaccination				<0.001 *
<6 months	255 (8.7)	60 (23.5)	18.6–29.0	
6–12 months	1245 (42.3)	482 (38.7)	36.0–41.4	
12–24 months	1260 (42.8)	586 (46.5)	43.8–49.3	
≥24 months	182 (6.2)	114 (62.6)	55.5–69.4	
Self-reported long COVID				0.031 *
No	792 (26.9)	360 (45.5)	42.0–48.9	
Yes	2150 (73.1)	882 (41.0)	39.0–43.1	
Pandemic fatigue				<0.001 *
Low	1196 (40.7)	290 (24.2)	21.9–26.7	
Moderate	1166 (39.6)	560 (48.0)	45.2–50.9	
High	580 (19.7)	392 (67.6)	63.7–71.3	

* A *p*-value less than 0.05 is considered to be statistically significant; † location" included urban areas (defined as main urban areas, urban–rural junction, and peri-urban areas) and rural areas (defined as townships and villages); "high school and below" included high school and below, technical secondary school, junior college and undergraduate students; master candidates were also included in the "Master's degree". For "relationship status", we divided participants into two categories based on the presence or absence of a lover or cohabiting spouse, with married but separated being considered unpartnered.

3.2. Pandemic Fatigue in the Post-Pandemic Era

As is shown in Figure 1, the average score of 2942 participants in PFS was 21.67 ± 8.86. The vaccine-hesitant group obtained a fatigue score of 25.26 ± 8.52, while that of the vaccine-accepting group was 19.04 ± 8.17. Disparities between the two groups were statistically significant ($p < 0.05$), including three information fatigue items and three behavioral fatigue items. The scores of six items in the vaccine-hesitant group were all significantly higher than those in the vaccine-accepting group.

Table 2 shows the characteristics across three pandemic fatigue groups. Only 1196 (40.7%) respondents reported a low level of pandemic fatigue. A higher level of pandemic fatigue clustered in people who are male, living in urban areas, smoking, drinking, struggling with chronic disease, higher perceived susceptibility, longer time since the most recent vaccination, and have self-reported long COVID. Moreover, this fatigue is unevenly distributed among people with different characteristics, except for relationship status and smoking history.

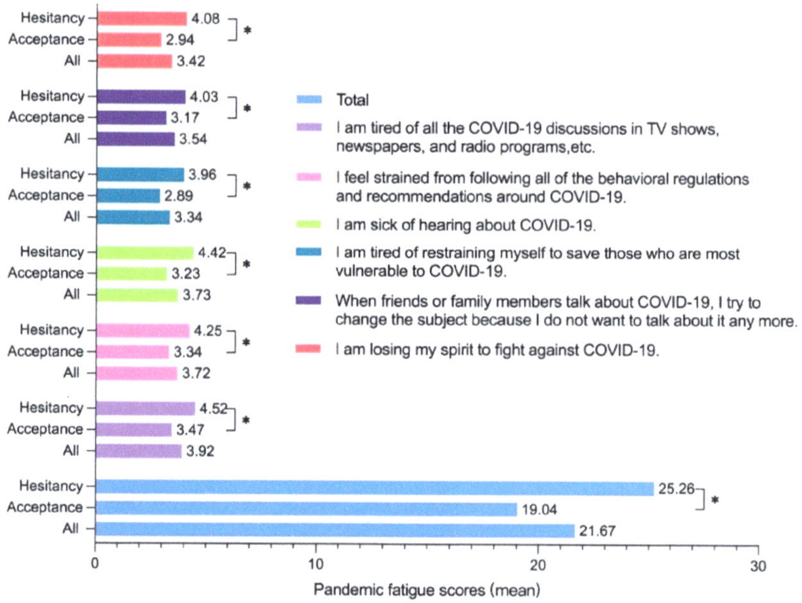

Figure 1. Pandemic Fatigue Scale item scores. * A *p*-value less than 0.05 is considered to be statistically significant.

Table 2. Pandemic fatigue characteristics of 2942 participants recovered from COVID-19 in China.

Characteristics †	Pandemic Fatigue (M ± SD)	*p* Value	Low Pandemic Fatigue	Moderate Pandemic Fatigue	High Pandemic Fatigue	*p* Value
			n (%)	n (%)	n (%)	
Sex		<0.001 *				0.007 *
Male	22.58 ± 9.01		448 (37.8)	475 (40.1)	262 (22.1)	
Female	21.05 ± 8.71		748 (42.6)	691 (39.3)	318 (18.1)	
Age (years)		0.447				0.019 *
<30	21.68 ± 8.68		538 (38.7)	598 (43.0)	254 (18.3)	
30–34	21.93 ± 8.99		324 (41.0)	291 (36.8)	175 (22.2)	
35–39	21.10 ± 9.18		189 (44.9)	149 (35.4)	83 (19.7)	
≥40	21.73 ± 8.93		145 (42.5)	128 (37.5)	68 (19.9)	
Location		0.004 *				0.043 *
Urban	21.81 ± 8.90		1067 (40.1)	1056 (39.7)	539 (20.2)	
Rural	20.28 ± 8.39		129 (46.1)	110 (39.3)	41 (14.6)	
Education		0.160				0.030 *
High school and below	21.26 ± 8.74		306 (40.0)	330 (43.1)	129 (16.9)	
Bachelor's degree	21.73 ± 8.90		795 (41.2)	744 (38.6)	389 (20.2)	
Master's degree	22.45 ± 8.93		95 (38.2)	92 (36.9)	62 (24.9)	
Relationship status		0.251				0.545
Without partner	21.31 ± 8.95		258 (41.1)	255 (40.7)	114 (18.2)	
With partner	21.77 ± 8.84		938 (40.5)	911 (39.4)	466 (20.1)	
Smoking		0.002 *				0.060
No	21.45 ± 8.84		1035 (41.5)	979 (39.3)	479 (19.2)	
Yes	22.86 ± 8.89		161 (35.9)	187 (41.6)	101 (22.5)	
Drinking		<0.001 *				0.001 *
No	21.00 ± 8.80		712 (42.9)	658 (39.6)	290 (17.5)	
Yes	22.53 ± 8.88		484 (37.8)	508 (39.6)	290 (22.6)	
Chronic disease		<0.001 *				<0.001 *
No	21.14 ± 8.82		837 (42.8)	780 (39.9)	337 (17.2)	
Yes	22.72 ± 8.85		359 (36.3)	386 (39.1)	243 (24.6)	
Perceived susceptibility		<0.001 *				<0.001 *
Low	21.01 ± 8.97		594 (43.7)	512 (37.7)	252 (18.6)	
Moderate	21.78 ± 8.56		474 (39.4)	509 (42.3)	221 (18.4)	
High	23.67 ± 9.14		128 (33.7)	145 (38.2)	107 (28.2)	

Table 2. *Cont.*

Characteristics †	Pandemic Fatigue (M ± SD)	*p* Value	Low Pandemic Fatigue	Moderate Pandemic Fatigue	High Pandemic Fatigue	*p* Value
			n (%)	n (%)	n (%)	
Perceived severity		0.944				0.011*
Low	21.63 ± 9.17		692 (41.8)	616 (37.2)	349 (21.1)	
Moderate	21.69 ± 8.38		385 (38.5)	440 (44.0)	175 (17.5)	
High	21.81 ± 8.73		119 (41.8)	110 (38.6)	56 (19.6)	
Time of the most recent vaccination		<0.001 *				0.001 *
<6 months	19.80 ± 8.99		129 (50.6)	87 (34.1)	39 (15.3)	
6–12 months	21.62 ± 8.74		510 (41.0)	495 (39.8)	240 (19.3)	
12–24 months	21.78 ± 8.87		505 (40.1)	502 (39.8)	253 (20.1)	
≥24 months	23.83 ± 9.03		52 (28.6)	82 (45.1)	48 (26.4)	
Self-reported long COVID		<0.001 *				<0.001 *
No	20.34 ± 9.05		386 (48.7)	266 (33.6)	140 (17.7)	
Yes	22.16 ± 8.75		810 (37.7)	900 (41.9)	440 (20.5)	
Vaccine hesitancy		<0.001 *				<0.001 *
No	19.04 ± 8.17		906 (53.3)	606 (35.6)	188 (11.1)	
Yes	25.26 ± 8.52		290 (23.3)	560 (45.1)	392 (31.6)	

* A *p*-value less than 0.05 is considered to be statistically significant; † "location" included urban areas (defined as main urban areas, urban–rural junctions, and peri-urban areas) and rural areas (defined as townships and villages); "high school and below" included high school and below, technical secondary school, junior college and undergraduate student; master candidates were also included in the "Master's degree". For "relationship status", we divided participants into two categories based on the presence or absence of a lover or cohabiting spouse, with married but separated being considered unpartnered.

3.3. *Vaccine Hesitancy toward the Next Dose of COVID-19 Vaccines*

Table 1 and Figure 2 demonstrate the vaccine hesitancy toward the next dose of COVID-19 vaccines. Of the 2942 participants, 1242 (42.2%) were hesitant (unwilling or not sure) to receive the next dose of COVID-19 vaccines, if available. A total of 835 (28.3%) participants marked "not sure" as the answer to the question about the intention to vaccinate against COVID-19, which accounted for 67.2% of the vaccine-hesitant group. As is shown in Table 1, among people with a history of COVID-19, those who are single, non-smokers, have low perceived severity after reinfection, did not report long COVID symptoms, and have a longer time since their last COVID-19 vaccination were more likely to be vaccine hesitant ($p < 0.05$). As the level of pandemic fatigue increased among the population, vaccine hesitancy also increased significantly. There was no significant difference in vaccination intention among people grouped by sex, age, location, education, drinking habits, chronic disease history, and perceived susceptibility to COVID-19 reinfection.

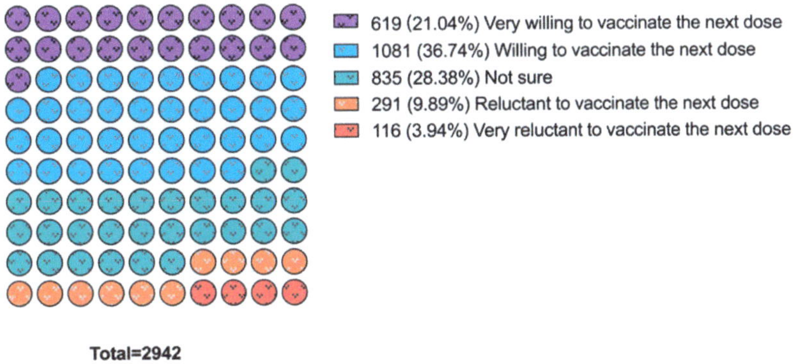

Figure 2. Willingness to receive the next dose of COVID-19 vaccines in recovered people (n = 2942).

3.4. Association between Pandemic Fatigue and Vaccine Hesitancy toward the Next Dose of COVID-19 Vaccines

The associations between pandemic fatigue and hesitancy toward the next dose of COVID-19 vaccines among people who have recovered from COVID-19 remained stable in a series of models (Table 3). In model A, controlling for no covariates, perceived moderate (cOR = 2.89, 95% CI: 2.42–3.44) or high (cOR = 6.51, 95% CI: 5.24–8.11) pandemic fatigue among people with a history of COVID-19 infection may lead to higher vaccine hesitancy. After adjusting for sociodemographic characteristics (gender, age, location, and education) in model B, the positive association between pandemic fatigue and vaccine hesitancy remained significant (moderate: aOR = 2.91, 95% CI: 2.44–3.47; high: aOR = 6.56, 95% CI: 5.27–8.17). We obtained similar results after adding health factors (drinking and chronic disease history) in multivariable logistic regression model C. The perceived susceptibility and severity of COVID-19 reinfection, which has been shown to affect vaccination intention in previous studies [27,28], were included in model D. The results showed that people with moderate and high levels of pandemic fatigue were 2.92 (95%CI: 2.45,-3.49) times and 6.80 (95%CI: 5.44–8.50) times more hesitant to receive the next dose of vaccine than those with low levels of fatigue, respectively. Furthermore, we adjusted for all covariates that were significantly unevenly distributed in pandemic fatigue in model E, in which the higher the level of pandemic fatigue among people with a history of COVID-19 infection, the more likely they were to be unwilling to receive the next dose of the COVID-19 vaccine (moderate: aOR = 2.94, 95% CI: 2.46–3.53; high: aOR = 6.88, 95% CI: 5.49–8.64).

Table 3. ORs (95%CI) for the hesitancy toward the next dose of COVID-19 vaccines according to pandemic fatigue.

Models	Low Pandemic Fatigue	Moderate Pandemic Fatigue	High Pandemic Fatigue
	OR (95%CI)	OR (95%CI)	OR (95%CI)
Model A	1 (Reference)	2.89 (2.42, 3.44) *	6.51 (5.24, 8.11) *
Model B	1 (Reference)	2.91 (2.44, 3.47) *	6.56 (5.27, 8.17) *
Model C	1 (Reference)	2.96 (2.48, 3.53) *	6.79 (5.44, 8.48) *
Model D	1 (Reference)	2.92 (2.45, 3.49) *	6.80 (5.44, 8.50) *
Model E	1 (Reference)	2.94 (2.46, 3.53) *	6.88 (5.49, 8.64) *

* A p-value less than 0.05 is considered to be statistically significant. Model A: unadjusted. Model B: adjusted for sociodemographic factors (gender, age, location, and education). Model C: model B plus health factors (drinking and chronic disease history). Model D: model C plus risk perception factors (perceived susceptibility and perceived severity). Model E: model D plus COVID-19 vaccination history and self-reported long COVID.

3.5. Subgroup Analyses

Subgroup analyses were exhibited in Table S2, and no modification was found in most subgroups (all p for interaction > 0.05), except for people stratified by education and smoking history. The association between pandemic fatigue and vaccine hesitancy weakened gradually with increasing education levels but remained significant. In the subgroup with the lowest education level, people who perceived high pandemic fatigue were 10.97 times more likely to be vaccine-hesitant than those with low pandemic fatigue, while it decreased to 3.88 times in the Master's degree subgroup. Furthermore, the impact of pandemic fatigue on vaccination intention was more pronounced in the smoking subgroup (moderate: aOR = 3.57, 95% CI: 2.09–6.09; high: aOR = 10.47, 95% CI: 5.60–19.57) than in the non-smoking subgroup (moderate: aOR = 2.94, 95% CI: 2.42–3.57; high: aOR = 6.60, 95% CI: 5.15–8.45).

4. Discussion

China experienced a nationwide Omicron variant outbreak from late 2022 to early 2023 [3,4]. To our knowledge, this is the first study to explore the COVID-19 vaccination intention and pandemic fatigue among people with a history of COVID-19 nearly six months after that event. According to our results, of the 2942 participants, 1242 (42.2%)

were hesitant to receive the next dose of COVID-19 vaccines, and 67.2% of the vaccine-hesitant participants were marked as "not sure". Only 40.7% of respondents reported a low level of pandemic fatigue, and the scores of all six pandemic fatigue items in the vaccine-hesitant group were significantly higher than those in the vaccine-accepting group. Based on the results of a series of regression models, people with moderate and high levels of pandemic fatigue were more likely to be hesitant to receive the next dose of COVID-19 vaccines than those with low levels of fatigue. These findings will help government authorities and relevant parties to take the potential threat behind the pandemic fatigue and vaccine hesitancy seriously, and promote future vaccination policies among infected people in an orderly and precise manner, thus ensuring the health and well-being of the population to the greatest extent possible.

In contrast to high COVID-19 vaccine acceptance in the general population, our findings found that 42.2% of the 2942 participants were hesitant to receive the next dose of COVID-19 vaccines [27,29–32]. A growing number of studies have demonstrated that patients who have recovered from COVID-19 have lower vaccination coverage and vaccine intention than the general population [8,13,33]. The Household Pulse Survey in the United States, which collected information on vaccination status (at least one dose) and vaccination intention of 63,266 people from 21 July to 2 August 2021, found that people with previously diagnosed COVID-19 had lower vaccine coverage (aPR = 0.88, 95%CI: 0.86–0.91) [13]. In terms of willingness to be vaccinated, people with a history of COVID-19 were more likely to be unwilling to receive the next dose of vaccine than uninfected people [13]. Gerussi et al. surveyed COVID-19 vaccine hesitancy in Italian patients who had recovered from COVID-19 in 2021 and found that 34.2% and 24.9%, respectively, were undecided or unwilling to receive COVID-19 vaccines [33]. In Wuhan, China, 1422 recovered patients had a 37.8% rate of COVID-19 vaccine hesitancy, since they believed they already had enough antibodies [8]. In addition, according to our results, those who are single, non-smokers, have low perceived severity after reinfection and did not report long COVID symptoms were more likely to be vaccine hesitant. Previous studies have also confirmed the relatively lower rate of hesitancy among people with more severe illness, high perceived severity, and the perception that they may develop sequelae [8,33,34]. These findings point to the need to focus on educating and confidence-building interventions for adults at the time of COVID-19 diagnosis, at clinic visits, or hospital discharge, as well as to better educate the public about the value of vaccination.

The direct and indirect effects of COVID-19 have caused great stress to the population and led to poor mental health outcomes [35]. According to our study, more than three years after COVID-19 raged, all participants showed varying degrees of pandemic fatigue, including information fatigue and health protection-related behavioral fatigue. Compared with a survey conducted in China in early 2022 [16], the average scores of the six items of PFS in our study were significantly higher, which may be related to the different survey times and respondents, but may also imply more obvious fatigue and reduced cooperation for epidemic prevention and control at present. In this study, a higher level of pandemic fatigue clustered in people who are male, living in urban areas, smoking, drinking, struggling with chronic disease, higher perceived susceptibility, longer time since the most recent vaccination, and have self-reported long COVID. Such an unexpected, prolonged pandemic will further worsen people's mental health and ability to cope with the situation, thus leading to mental fatigue, which is thought to be a natural psychological response of individuals through overexposure to negative information related to the pandemic and repeated implementation of behavioral restrictions [36]. On 5 May 2023, the World Health Organization declared that COVID-19 no longer constituted a PHEIC, but had transformed into an established and ongoing global health problem. Many people have generally negative attitudes toward chronic public health crises and have less motivation to engage in protective behaviors [36]. Yue et al. investigated fatigue during the three waves of the epidemic in China and showed that with the development of the epidemic, people experienced different degrees of pandemic fatigue, which may affect the occurrence of psychosomatic symptoms and perceived stress, reminding government authorities to pay attention to this phenomenon to avoid the occurrence of potential crises [36].

Across a range of models, pandemic fatigue was consistently positively associated with vaccine hesitation rates among those recovering from COVID-19, which became more significant with increasing adjustment. Previous studies in different populations have also reached similar conclusions [18,37]. Ali-Saleh et al. surveyed 2843 Arab parents and found that pandemic fatigue was indirectly associated with parents' less positive attitudes toward vaccinating their children [37]. An anonymous cross-sectional study of general adults in Malaysia found that those with a lower pandemic fatigue score were more willing to be vaccinated against COVID-19 (OR = 2.34, 95% CI: 1.75–3.22) [18]. Pandemic fatigue is an expected and natural response to a prolonged public health crisis [18,38]. The concept of behavioral fatigue related to compliance with COVID-19 restrictions or pandemic fatigue is a social issue. During this pandemic, the impact of mental fatigue has started to have a cascading impact on vaccination efforts [39]. Fatigue may begin to cast doubt on the effectiveness of COVID-19 mitigation strategies and reluctance to take steps to end the pandemic, including vaccination [18]. Thus, pandemic fatigue is a potential correlate of vaccine acceptance and may impede the translation of vaccination intentions into behaviors. Addressing pandemic fatigue requires a robust multipronged response that addresses motivation in terms of the costs and benefits of mitigation [18].

This is a nationwide study encompassing all provinces in mainland China, which is representative and statistically efficacious. However, as described in the Methods section, this is a cross-sectional study and caution is needed when making causal inferences. Second, only those with access to the Internet were able to participate in this survey, which to some extent ignores the remaining portion of the population and may lead to selection bias. Third, we did not adjust for social media use, income, occupation, and vaccine production information when analyzing the association between pandemic fatigue and vaccine hesitancy, and there is a possibility that the level of association was misestimated. Additionally, it is important to note that this study was conducted only in patients who had recovered from COVID-19, so the levels of pandemic fatigue and vaccine hesitancy in this study were not representative of the general population. Future studies with longitudinal or experimental designs are encouraged to confirm our findings and elucidate potential mechanisms and interventions to reduce pandemic fatigue and vaccine hesitancy.

5. Conclusions

In this study, 42.2% of participants with a history of COVID-19 were hesitant to receive the next dose of COVID-19 vaccines, and only 40.7% of respondents reported a low level of pandemic fatigue. People with higher levels of pandemic fatigue were more likely to be hesitant to receive the next dose of COVID-19 vaccines. Therefore, given the ongoing reinfections, implementing a health education plan to reduce pandemic fatigue and prioritizing vaccination issues for people with a history of COVID-19 may be vital in promoting the reduction of the COVID-19 disease burden and ensuring the health and well-being of the population.

Supplementary Materials: The following supporting information can be downloaded at: https://www.mdpi.com/article/10.3390/vaccines11101570/s1, Table S1: collection of valid questionnaires by region in mainland China; Table S2: subgroup analysis of the association between pandemic fatigue and COVID-19 vaccine hesitancy among 2942 participants who have recovered from COVID-19 in China.

Author Contributions: Conceptualization, C.Q., J.D. and J.L.; methodology and analysis, C.Q. and J.L.; visualization, C.Q. and W.Y.; writing—original draft preparation, C.Q.; writing—review and editing, M.D., Y.W. and Q.L.; supervision, M.L. and J.L. All authors have read and agreed to the published version of the manuscript.

Funding: This study was funded by the the Beijing Natural Science Foundation (L222027). The funders had no role in the study design, data collection and analysis, decision to publish, or preparation of the paper. No payment was received by any of the co-authors for the preparation of this article.

Institutional Review Board Statement: This study met the requirements of the Declaration of Helsinki and was approved by the Ethics Committee of Peking University (IRB00001052-21126).

Informed Consent Statement: Informed consent was obtained from all subjects involved in the study.

Data Availability Statement: All data in the study are available from the corresponding author by request.

Acknowledgments: We would like to express our sincere gratitude to all the participants who enrolled in this study.

Conflicts of Interest: The authors declare no conflict of interest.

References

1. WHO. COVID-19 Variants. 2023. Available online: https://www.who.int/activities/tracking-SARS-CoV-2-variants (accessed on 23 August 2023).
2. WHO. COVID-19 Dashboard. 2023. Available online: https://www.who.int/emergencies/diseases/novel-coronavirus-2019 (accessed on 23 August 2023).
3. The Joint Prevention and Control Mechanism of the State Council. Press Conference of the State Council Joint Prevention and Control Mechanism on 8 May 2023. 2023. Available online: http://www.nhc.gov.cn/xwzb/webcontroller.do?titleSeq=11516&gecstype=1 (accessed on 23 August 2023).
4. Fu, D.; He, G.; Li, H.; Tan, H.; Ji, X.; Lin, Z.; Hu, J.; Liu, T.; Xiao, J.; Liang, X.; et al. Effectiveness of COVID-19 Vaccination Against SARS-CoV-2 Omicron Variant Infection and Symptoms—China, December 2022–February 2023. *China CDC Wkly.* **2023**, *5*, 369–373. [CrossRef] [PubMed]
5. WHO. Statement on the Fifteenth Meeting of the IHR (2005) Emergency Committee on the COVID-19 Pandemic. 2023. Available online: https://www.who.int/news/item/05-05-2023-statement-on-the-fifteenth-meeting-of-the-international-health-regulations-(2005)-emergency-committee-regarding-the-coronavirus-disease-(covid-19)-pandemic (accessed on 23 August 2023).
6. Haynes, B.F.; Corey, L.; Fernandes, P.; Gilbert, P.B.; Hotez, P.J.; Rao, S.; Santos, M.R.; Schuitemaker, H.; Watson, M.; Arvin, A. Prospects for a safe COVID-19 vaccine. *Sci. Transl. Med.* **2020**, *12*, eabe0948. [CrossRef] [PubMed]
7. Dubé, E.; Laberge, C.; Guay, M.; Bramadat, P.; Roy, R.; Bettinger, J.A. Vaccine hesitancy: An overview. *Hum. Vaccines Immunother.* **2013**, *9*, 1763–1773. [CrossRef] [PubMed]
8. Huang, Y.; Zhang, L.; Fu, J.; Wu, Y.; Wang, H.; Xiao, W.; Xin, Y.; Dai, Z.; Si, M.; Chen, X.; et al. COVID-19 Vaccine Hesitancy Among Patients Recovered From COVID-19 Infection in Wuhan, China: Cross-Sectional Questionnaire Study. *JMIR Public Health Surveill.* **2023**, *9*, e42958. [CrossRef]
9. WHO. Ten Threats to Global Health in 2019. 2020. Available online: https://www.who.int/news-room/spotlight/ten-threats-to-global-health-in-2019 (accessed on 23 August 2023).
10. Nuwarda, R.F.; Ramzan, I.; Weekes, L.; Kayser, V. Vaccine Hesitancy: Contemporary Issues and Historical Background. *Vaccines* **2022**, *10*, 1595. [CrossRef]
11. Letizia, A.G.; Ge, Y.; Vangeti, S.; Goforth, C.; Weir, D.L.; Kuzmina, N.A.; Balinsky, C.A.; Chen, H.W.; Ewing, D.; Soares-Schanoski, A.; et al. SARS-CoV-2 seropositivity and subsequent infection risk in healthy young adults: A prospective cohort study. *Lancet Respir. Med.* **2021**, *9*, 712–720. [CrossRef]
12. El-Ghitany, E.M.; Ashour, A.; Omran, E.A.; Farghaly, A.G.; Hassaan, M.A.; Azzam, N.F.A.E.-M. COVID-19 vaccine acceptance rates and predictors among the Egyptian general population and Healthcare workers, the intersectionality of age and other factors. *Sci. Rep.* **2022**, *12*, 19832. [CrossRef]
13. Nguyen, K.H.; Huang, J.; Mansfield, K.; Corlin, L.; Allen, J.D. COVID-19 Vaccination Coverage, Behaviors, and Intentions among Adults with Previous Diagnosis, United States. *Emerg. Infect. Dis.* **2022**, *28*, 631–638. [CrossRef]
14. Bodas, M.; Kaim, A.; Velan, B.; Ziv, A.; Jaffe, E.; Adini, B. Overcoming the effect of pandemic fatigue on vaccine hesitancy—Will belief in science triumph? *J. Nurs. Sch.* **2023**, *55*, 262–271. [CrossRef]
15. Al-Tammemi, A.B.; Tarhini, Z.; Akour, A. A swaying between successive pandemic waves and pandemic fatigue: Where does Jordan stand? *Ann. Med. Surg.* **2021**, *65*, 102298. [CrossRef]
16. Xin, L.; Wang, L.; Cao, X.; Tian, Y.; Yang, Y.; Wang, K.; Kang, Z.; Zhao, M.; Feng, C.; Wang, X.; et al. Prevalence and influencing factors of pandemic fatigue among Chinese public in Xi'an city during COVID-19 new normal: A cross-sectional study. *Front. Public Health* **2022**, *10*, 971115. [CrossRef]
17. Asimakopoulou, E.; Paoullis, P.; Shegani, A.; Argyriadis, A.; Argyriadi, A.; Patelarou, E.; Patelarou, A. Translation, Adaptation and Validation of the Pandemic Fatigue Scale (PFS) in the Greek Language. *Healthcare* **2022**, *10*, 2118. [CrossRef] [PubMed]
18. Wong, L.P.; Alias, H.; Siaw, Y.-L.; Muslimin, M.; Lai, L.L.; Lin, Y.; Hu, Z. Intention to receive a COVID-19 vaccine booster dose and associated factors in Malaysia. *Hum. Vaccines Immunother.* **2022**, *18*, 2078634. [CrossRef] [PubMed]

19. Qin, C.; Wang, R.; Tao, L.; Liu, M.; Liu, J. Association Between Risk Perception and Acceptance for a Booster Dose of COVID-19 Vaccine to Children Among Child Caregivers in China. *Front. Public Health* **2022**, *10*, 834572. [CrossRef] [PubMed]
20. Qin, C.; Yan, W.; Du, M.; Liu, Q.; Tao, L.; Liu, M.; Liu, J. Acceptance of the COVID-19 vaccine booster dose and associated factors among the elderly in China based on the health belief model (HBM): A national cross-sectional study. *Front. Public Health* **2022**, *10*, 986916. [CrossRef]
21. Daud, S. The COVID-19 Pandemic Crisis in Malaysia and the Social Protection Program. *J. Dev. Soc.* **2021**, *37*, 480–501. [CrossRef]
22. Star Survey. Sample Service. 2023. Available online: https://www.wjx.cn/sample/service.aspx (accessed on 23 August 2023).
23. Qin, C.; Du, M.; Wang, Y.; Liu, Q.; Yan, W.; Tao, L.; Liu, M.; Liu, J. Assessing acceptability of the fourth dose against COVID-19 among Chinese adults: A population-based survey. *Hum. Vaccines Immunother.* **2023**, *19*, 2186108. [CrossRef]
24. Lilleholt, L.; Zettler, I.; Betsch, C.; Böhm, R. Pandemic fatigue: Measurement, correlates, and consequences. *PsyArXiv* **2020**, 1–36. [CrossRef]
25. Marsella, A.J.; Leong, F.T.L. Cross-Cultural Issues in Personality and Career Assessment. *J. Career Assess.* **1995**, *3*, 202–218. [CrossRef]
26. Janz, N.K.; Becker, M.H. The Health Belief Model: A Decade Later. *Health Educ. Q.* **1984**, *11*, 1–47. [CrossRef]
27. Qin, C.; Wang, R.; Tao, L.; Liu, M.; Liu, J. Acceptance of a Third Dose of COVID-19 Vaccine and Associated Factors in China Based on Health Belief Model: A National Cross-Sectional Study. *Vaccines* **2022**, *10*, 89. [CrossRef]
28. Baccolini, V.; Renzi, E.; Isonne, C.; Migliara, G.; Massimi, A.; De Vito, C.; Marzuillo, C.; Villari, P. COVID-19 Vaccine Hesitancy among Italian University Students: A Cross-Sectional Survey during the First Months of the Vaccination Campaign. *Vaccines* **2021**, *9*, 1292. [CrossRef]
29. Liu, Y.; Han, J.; Li, X.; Chen, D.; Zhao, X.; Qiu, Y.; Zhang, L.; Xiao, J.; Li, B.; Zhao, H. COVID-19 Vaccination in People Living with HIV (PLWH) in China: A Cross Sectional Study of Vaccine Hesitancy, Safety, and Immunogenicity. *Vaccines* **2021**, *9*, 1458. [CrossRef]
30. Hong, J.; Xu, X.-W.; Yang, J.; Zheng, J.; Dai, S.-M.; Zhou, J.; Zhang, Q.-M.; Ruan, Y.; Ling, C.-Q. Knowledge about, attitude and acceptance towards, and predictors of intention to receive the COVID-19 vaccine among cancer patients in Eastern China: A cross-sectional survey. *J. Integr. Med.* **2022**, *20*, 34–44. [CrossRef]
31. Hu, S.; Liu, J.; Li, S.; Wu, Q.; Wang, X.; Xu, D.; Chen, Y. Patients with IBD have a more cautious attitude towards COVID-19 vaccination. *Front. Immunol.* **2023**, *13*, 1077308. [CrossRef]
32. Qin, C.; Yan, W.; Tao, L.; Liu, M.; Liu, J. The Association between Risk Perception and Hesitancy toward the Booster Dose of COVID-19 Vaccine among People Aged 60 Years and Older in China. *Vaccines* **2022**, *10*, 1112. [CrossRef]
33. Gerussi, V.; Peghin, M.; Palese, A.; Bressan, V.; Visintini, E.; Bontempo, G.; Graziano, E.; De Martino, M.; Isola, M.; Tascini, C. Vaccine Hesitancy among Italian Patients Recovered from COVID-19 Infection towards Influenza and Sars-Cov-2 Vaccination. *Vaccines* **2021**, *9*, 172. [CrossRef] [PubMed]
34. Petravić, L.; Arh, R.; Gabrovec, T.; Jazbec, L.; Rupčić, N.; Starešinič, N.; Zorman, L.; Pretnar, A.; Srakar, A.; Zwitter, M.; et al. Factors Affecting Attitudes towards COVID-19 Vaccination: An Online Survey in Slovenia. *Vaccines* **2021**, *9*, 247. [CrossRef] [PubMed]
35. Wu, T.; Jia, X.; Shi, H.; Niu, J.; Yin, X.; Xie, J.; Wang, X. Prevalence of mental health problems during the COVID-19 pandemic: A systematic review and meta-analysis. *J. Affect. Disord.* **2021**, *281*, 91–98. [CrossRef] [PubMed]
36. Yue, Y.; Li, L.; Liu, R.; Zhang, Y.; Zhang, S.; Sang, H.; Tang, M.; Zou, T.; Shah, S.M.; Shen, X.; et al. The dynamic changes of psychosomatic symptoms in three waves of COVID-19 outbreak and fatigue caused by enduring pandemic in China. *J. Affect. Disord.* **2023**, *331*, 17–24. [CrossRef] [PubMed]
37. Ali-Saleh, O.; Bord, S.; Basis, F. Factors Associated with Decisions of Arab Minority Parents in Israel to Vaccinate Their Children against COVID-19. *Vaccines* **2022**, *10*, 870. [CrossRef] [PubMed]
38. WHO. Pandemic Fatigue: Reinvigorating the Public to Prevent COVID-19. 2020. Available online: https://apps.who.int/iris/bitstream/handle/10665/335820/WHO-EURO-2020-1160-40906-55390-eng.pdf (accessed on 23 August 2023).
39. Reicher, S.; Drury, J. Pandemic fatigue? How adherence to covid-19 regulations has been misrepresented and why it matters. *BMJ* **2021**, *372*, n137. [CrossRef] [PubMed]

MDPI AG
Grosspeteranlage 5
4052 Basel
Switzerland
Tel.: +41 61 683 77 34

Vaccines Editorial Office
E-mail: vaccines@mdpi.com
www.mdpi.com/journal/vaccines

www.ingramcontent.com/pod-product-compliance
Lightning Source LLC
Chambersburg PA
CBHW042010260926
51850CB00020B/2965
9783725826469